£5.95

RALACHA
PASS
TOPOGOGMA

TOPOYOGMA

DUTUNG

RATAL   LOSAR   HANSA   PARANG

PASS

S   P   I   T   I

ACHU

MORANG

PHUTI RUNI

GAZA

LARA

CHABRANG

DANKHAR

KARN

KULING

Spiti R.

ULU   BULDUR

K U N A W A R

Sutlej River

RAMPUR

KUMHAR
SAIN

MATIANA

THEOG

HIMALAYAN
CIRCUIT

The Story of a Journey in the
Inner Himalayas

HIMALAYAN CIRCUIT

The Story of a Journey in the
Inner Himalayas

by

G. D. KHOSLA

with a Foreword by

JAWAHARLAL NEHRU

DELHI
OXFORD UNIVERSITY PRESS
BOMBAY CALCUTTA MADRAS

Oxford University Press, Walton Street, Oxford OX2 6DP

New York Toronto
Delhi Bombay Calcutta Madras Karachi
Petaling Jaya Singapore Hong Kong Tokyo
Nairobi Dar es Salaam
Melbourne Auckland
and associates in
Berlin Ibadan

First published by Macmillan & Co Ltd.,
London, UK, 1956

First Indian impression in
Oxford India Paperbacks 1989

Printed at Rekha Printers Pvt. Ltd., New Delhi 110020
and published by S. K. Mookerjee, Oxford University Press
YMCA Library Building, Jai Singh Road, New Delhi 110001

FOREWORD

BY

JAWAHARLAL NEHRU

WHAT is it that draws men to the high mountains and to the deserts? What urge leads them to write about them? There are many books about mountain treks and desert journeys and some of them are classics of their kind. And yet there are not enough of these books and a new one is always welcome.

I do not know much about deserts except that my little acquaintance with them has rather fascinated me. But mountains, and especially the higher altitudes where one can commune with the snowy peaks, attract me enormously. With the Himalayas I feel a little intimate, partly because I have seen much more of them, partly because they are so wrapped up in India's story and legend. Once, thirteen years ago, I went up the Kulu valley to Manali. That was as far as I could go then as I had to be back to the hot plains below. But from Manali I looked in the direction of the high passes beyond which lie Spiti and Lahoul. I consoled myself with the thought that I would come again and cross that mountain barrier to this new land which was so utterly different from the India I knew.

But that opportunity has not come to me, and I fear it will never come now. New Delhi, with its

v

strange and rather unreal atmosphere and its multi-
farious occupations, holds me prisoner, and even when
I go to the mountains, it is to some place of relatively
easy access and for two or three days only. I fear there
are no long treks for me now, and perhaps I have passed
the age for these feats of valour. This is a disturbing
thought.

I remember many a trek in these solitudes of the
Himalayas. Most of them were in the upper valleys
of Kashmir, which exercise such a powerful fascination
on me as on many others. I remember the wonderful
lakes edged by some glacier and often partly covered
with a layer of ice. In the pride of my younger days,
and perhaps to show off a little, I jumped into them
and swam about a little, almost frozen in the process.
I remember also, with a feeling of nostalgia, wonderful carpets of flowers right up to the edge of the snow,
and the bracing air which brought a new dimension to
life. There were few human beings at those altitudes,
but there was enchantment in that loneliness, a sense of
vast spaces and something of eternity.

Not having been able to go to Spiti and Lahoul, I
have read this book to get some second-hand knowledge of this area. I hope the book will bring some
breath of the inner Himalayas and of a strange land to
the unfortunate people who always live in the plains
below and know little of the joys and risks and dangers
of the high mountains.

J. N.

PREFACE

THIS book does not contain stories of spectacular feats of mountaineering, nor is it a record of man's triumph over Nature. It is simply an account of a journey through a region of the Himalayas which deserves to be known better. It is the region which lies beyond the Great Himalayan Range, near the Indo-Tibetan border on the north-west, between N. latitude 30° 8′ and 33°, and E. longitude 76° 49′ and 78° 35′. Here lie the valleys of Spiti and Lahoul at a height of 12,000 feet to 16,000 feet above sea-level, in the rugged grandeur and breath-taking beauty of the Inner Himalayas. The people inhabiting these valleys live a life of unspoilt isolation. Their strange attire, their Mongolian features and their unwashed bodies give them a *farouche* appearance, but they have a cheerful and kindly disposition. Their social economy, their customs and the peculiar form of Buddhism which they practise are stream-lined to meet the exigencies of their situation, and are calculated to reduce the resistance of the elements and soften the cruel impact of Nature. Life in Lahoul, and to a much greater extent in Spiti, is hard. Communications are unsatisfactory and, in some parts, non-existent. The amenities of modern life are wholly lacking. In Spiti there are few bridges. Streams and torrents must be forded, sometimes at considerable risk. There are hardly any schools, and in the whole valley there is not a single hospital, or a doctor to give medical aid. The mortality rate is high and the expectation of life low. I saw only one or two old people during the course of my visit, and very few appeared to have crossed the fifty-year mark.

I undertook this journey because the Himalayas have an overpowering fascination for me. Despite their seemingly cruel aspect, and their unfriendly behaviour towards visitors, they have continued to draw me year after year. Much of my childhood was spent in the lap of the outer Himalayas where the hill-stations and holiday resorts are situated. When I was ten, my father took me for my first trek into the interior. It was a brief eight-day affair, but I remember clearly the delights of my first contact with real mountains. I remember being continuously surprised and excited by new dis-coveries at every step — the fragrance of pines, the biting quality of the evening wind, the brightness of the midday sun, the deep blue of the sky, an overhanging rock with ferns and strange flowers painted across its face, a mountain peak that stood with its head above the clouds and caught the morning sun before it ever arose, a stream of water issuing from under a rock which seemed to have been placed there on purpose so that weary travellers could quench their thirst, and a thousand other wonders. The people that lived in these parts were equally strange and wonderful. During a lunar eclipse they came out with their guns and sticks and shot at Rahu, the demon who had caught the face of the moon in his unclean hands. They threw stones and other missiles at him and shouted as loudly as they could : 'Rahu, you miscreant, let go', till the moon was free from the obscurity of earth's shadow.

Except for a few years spent in Europe, I have gone to the Himalayas and lived among the mountains for a space of time each year. I have wandered in the wildernesses of the inner ranges and experienced the difficulties and hazards of Himalayan travel. I have been moved by the peace and beauty of an arcadian scene with soft green grass and a profusion of wild flowers all around, and awed by the rugged majesty of a gigantic monolith or the sight of an angry torrent rushing out of an ice cave. The Himalayas have something which no other mountains anywhere else have. It is not merely their size, though that alone is enough to stir the most phlegmatic observer. It is partly the superlative degree of every impact to which they subject the senses, and the extremes of sensuous experience which they provoke ; but the magic spell of the Himalayas is really woven by the gods and by the spirits of sages and saints who have dwelt among them from time immemorial.

I have never had a desire to conquer the Himalayas or climb virgin peaks. I have no wish to triumph over Nature and cover myself with glory. To me the Himalayas are neither friends nor foes. They are something essential to my emotional existence. I want to be near them whether they smile or frown. I do not know if they will protect me or destroy me, but I do know that nothing moves me to the same extent and in the same manner as the Himalayas.

It was to express this and something of what I feel every time I see these mountains and go close to them that I began to write this book. I do not know if I have succeeded in communicating even a small measure of the emotions which the Himalayas arouse in me, and if it was worth while making the effort ; but as the narrative proceeded, I lived again the moments of delight and ecstasy and of fear and wonder that I had experienced during the actual journey, and I felt that my labour was justified.

But there was one other reason. It was my desire to do something for the people of Spiti and Lahoul by advertising their plight and the conditions in which they exist. I have achieved a little by writing about these matters in the press and by discussing them with our public men. I hope this book will go some little way in bringing the issue to the forefront, so that a little more attention can be paid to this backward but magnificent region of our country.

CONTENTS

I
SIMLA TO KULU

HIM 584 was waiting on the Cart Road near the Cecil Hotel garages. A bunch of red-coated orderlies stood admiring its blue-green colour and its snub, determined snout. Among them was Sundar, my *jemadar*, wearing his new scarlet and gold livery. The Land Rover matched him in pride and brilliance. I hoped, in performance, it would prove as satisfactory. My roll of bedding and my steel trunk were already in. I found a safe corner for the bag containing my cameras and stepped into the seat beside the driver.

The Land Rover is a proud animal, with its hard over-inflated tyres and stiff unbending springs. HIM 584 was certainly no exception, and as the red-coats bowed and salaamed an obsequious farewell, I thought for a moment that the Land Rover was the real object of their servile attention.

We bumped along at a steady 20 m.p.h. and I felt there was very little I could complain about. I was going to see real mountains. I was going over and across the Great Himalayas and everything else was unimportant.

I tried to converse with the driver — a young Himachali of about twenty-five ; but he was a reluctant talker, and spoke only in monosyllables. I let him focus his attention on the tortuous road till we reached Theog, eighteen miles from Simla. Here a chain drawn across the road barred our way. The road between Theog and Narkanda is narrow, and only one-way traffic is allowed. The traffic in each direction moves between fixed times. I was told I should have to wait more than an hour, before I could be allowed to go forward. I had anticipated this ; and had taken the precaution to obtain a permit, entitling me to travel through, irrespective of the time-control regulations. With a little care one could drive in safety, for the road permitted the passage of two cars in many places.

I showed my permit to the constable on duty. He said he was illiterate, and had orders not to let anyone through before 5 P.M.

'Whose orders ?' I asked.

'The *thanedar* ¹ *sahib*'s.'

'And where is the *thanedar sahib* ?'

'In his house up there.'

I was beginning to lose my patience : 'Will you go and fetch him. Say he is wanted by . . .' and I mentioned my name and designation as pompously as I could.

The constable was away for five minutes. He returned alone, and said the *thanedar sahib* could not leave his post. I almost ran up the steep path leading to the dilapidated building which housed the *thanedar*'s office and residence.

A white-bearded Sikh sat on a charpoy, playing with a bundle of papers. He was turbanless and wore his hair done up in a diminutive bun, perched above his

¹ Sub-Inspector of Police, officer in charge of a police-station.

forehead. I thrust my permit in front of his nose, and asked him to let me proceed on my way. He examined the piece of paper for a long minute, and read and re-read the single line several times over. Then he cleared his throat very loudly and said :

'This permit allows you to go to Narkanda "against the timings" which means in accordance with the timings.'

I tried to explain the meaning of the word 'against'. I argued that nobody needed a permit to travel in accordance with the timings and the only reason I had taken the trouble to obtain the permit was . . .

But polite talk was useless ; for the *thanedar* 'against' clearly meant 'in accordance with' ; so I threatened to report him and told him what would happen if he deliberately flouted the orders of the Chief Commissioner, and unlawfully held up a High Court Judge. Slowly and very reluctantly he put down his papers, got up from his charpoy, and standing in front of his door, signalled acquiescence to the constable at the barrier.

The rest of our journey to Narkanda was uneventful. We met hardly any oncoming traffic, and since the road was broader and safer than I had anticipated there were no thrills or missed heart-beats. The driver remained as uncommunicative as before, and I began to feel that he must have a grievance against me. Perhaps he was a newly married man and I had taken him away from his wife. I asked him if he was married.

'No, sir,' he said and was silent again.

We reached Narkanda at half-past five and stopped for tea. I was lucky to find my friend Chaudri, the Inspector-General of Police, Himachal, and his wife at

B

the rest-house. They gave me tea, and laughed heartily at my story of the Theog *thanedar*.

At six I left Narkanda and started down the steep road which passes through the bazaar and winds its way down the thirteen miles to Luhri. I had been warned about this part of the road, and was told that the Land Rover would, in all probability, not go beyond Kumharsain, half way between Narkanda and Luhri. I decided to drive as far as I could and walk the rest of the way. My baggage would have to be brought down by the jeep which was waiting for me at Luhri.

The road beyond Kumharsain proceeds down a steep barren hill in a series of hairpin bends — 'scissors' the local people call them. The road is narrow and unprotected by a parapet. At each bend I had to get down and put stones in front of the wheels. The nature of the road demanded far more condescension than the stiff-necked Land Rover was capable of. Its restricted steering-lock did not permit a sharp turn in one quick movement, and we had to reverse the car each time.

My reticent companion broke his silence once or twice to abuse the road and the Land Rover, when the wheels failed to grip the loose surface of the unmetalled path. I thanked God for a dry day — a shower of rain would have played havoc with the clayey soil of the hill, and made it impossible to control anything on wheels proceeding along such a steep, narrow and winding pathway. The papers had predicted 'dry weather' for forty-eight hours beginning with Thursday morning and it was only Thursday afternoon. This would just see me through, and I blessed the weather clerks.

The hairpin bends became more and more trouble-some. We were travelling at a speed of less than 7 m.p.h.
I began to strain my eyes for each furlong stone. $4\frac{1}{2}$ miles to Luhri, $4\frac{3}{8}$ miles, $4\frac{1}{4}$, $4\frac{1}{8}$. I thought : 'I can walk the rest of the way easily and be home before nightfall.'
We could take the left-hand turns without much difficulty, but the right-hand turns became a series of nightmares. I suppose it had something to do with the rotation of the earth. At mile 49·4 we came to a whole bunch of 'scissors' zigzagging all the way down a steep slope. The driver stopped, wiped his brow and re-mained sitting in his seat.

I jumped out, and telling him to stay where he was, ran straight down the hill. In twenty minutes I was walking up the path to the rest-house. The *naib tahsildar* [1] and the driver of the jeep which had been sent from Kulu to fetch me were coming out to look for me. I explained the position to them and told the driver to take the jeep up and bring my baggage and the driver of the Land Rover. The driver said he would bring the Land Rover down. I didn't see how he was going to do it, but it seemed useless to argue the matter. From the rest-house I sent up three more men to help with the baggage.

Then began a long wait, for my dinner was in the Land Rover. Luhri lies at the bottom of the Sutlej valley, at a height of less than 2000 feet above sea-level. In midsummer the day temperatures rise high, though, in the evening, there is often a pleasant breeze. As I strolled up and down the level ground in front of the rest-house, I felt hot and discontented walking one way,

[1] Revenue official with miscellaneous administrative duties.

but when I turned in the opposite direction I met a cool breeze that took away some of the gloom and frustration of a long evening.

The moon came slowly up from behind the rest-house and the hill-tops opposite were covered with a ghostly mist. Three hundred feet below me the Sutlej roared, groaned, grumbled, chuckled or whatever else it was in the habit of doing.

Nine o'clock and still there was no sign of my baggage. I was beginning to feel really hungry. I hoped the driver of the jeep had not been so ill-advised as to try to bring the Land Rover down. If he had . . .

Suddenly a wide brush of light swept the bushes far up on the hillside. Could that be the Land Rover ? It was quite dark for a few seconds and again the bushes lit up for a moment.

The Land Rover came in slowly, like the early railway trains, with a man walking in front, and making frantic gestures to guide it along the narrow path. It had taken over two hours to travel a distance of two and a half miles, but it was safely home ; the driver was no longer scared and my dinner was soon warmed up and served.

The caretaker of the rest-house advised me to sleep inside. The Luhri mosquitoes, he said, were monsters of malice and destruction. They carried malarial germs of an inordinately vicious breed. I should be safe if I had my bed inside, for the wire-gauze doors refused entry to the mosquitoes. I soon found out that the caretaker's mosquitoes were really sand-flies and they were able to slip through the meshes of the protective gauze. Several hundreds of them began to buzz about my ears and dig needles into my face, hands and feet. The air inside the room was hot and oppressive. Sleep was quite impossible. If I went out, the sand-flies would continue to entertain me, but I should at least have fresh air and a cool breeze. So at midnight I got up and dragged my bed out and lay down under a bright moon. There were no sand-flies. The breeze was too much for them and I was soon sound asleep.

I woke up at 4.40 A.M. and saw that daylight filled the whole valley. I tried to sleep again, but soon found it was no use. So at five o'clock I got up, bathed and dressed and was ready to leave. We started at six in an old yellow jeep which looked and sounded as if it would go for a mile or two and then collapse and disintegrate. Half a mile beyond the Sutlej bridge we had to stop, unload and take the jeep at a snail's pace across a difficult portion of the road. The hillside had given way, and taken ten feet of the road with it. Workmen were busy making a crude form of what the engineers call a gallery bridge. Stakes of pine-wood driven into the hillside rested on a cross-beam and the road surface was made of branches and loose earth piled over the stakes. The gallery bridge was narrow and shaky. The outer wheels of the jeep were hardly two inches from the edge below which the Sutlej groaned and gurgled. Even so the rocks on the other side twice fouled the wings and there was a horrible scraping sound as the jeep crossed over into safety. The driver was quite cheerful about the whole thing and told me that as long as I was with him nothing could go wrong. For hadn't the Commissioner Sahib's jeep stopped at this very place just as the hillside collapsed two months ago? The gods protect the mighty, he added. I didn't think I was quite as mighty as that, but the driver's belief in the importance of the load he was carrying and the righteousness of his mission were a source of comfort and solace to me throughout the difficult — I shall not say perilous — journey to Kulu.

We reached Ani at a quarter to eight, covering the twelve miles from Luhri in an hour and forty-five minutes. Here the *tahsildar* was waiting for me and had breakfast ready. His cook had risen to a sweet dish — a kind of *firni* [1] which I found quite palatable. We were on our way again at half-past eight. The road became steadily worse and on either side of the Jalori pass (10,000 feet above sea-level) there were so many sharp turns, and the path was so steep and narrow, that the driver needed all his skill and all his belief in the protection which the gods afforded me on this route. The engine revolutions had to be kept high, for low running meant loss of power and consequent stalling of the engine ; bends had to be negotiated with a quick and decisive movement of the steering-wheel. A slight delay, a little hesitation, and the jeep would have gone hurtling over a 200-foot precipice, while a premature movement would have brought its nose crashing into the hillside.

We crossed the Jalori pass and reached Shoja (8800 feet above sea-level) — the first village in the Kulu valley along this road. The rest-house is situated above the village in a small clearing in the heart of a pine forest, and commands a magnificent view of the moun-tain ranges towards the north. One could sit for hours basking in the sun and gazing at the distant peaks. I

---

[1] Ground rice cooked in milk.

was not sorry when the driver reported slight engine trouble. I got out and filled my lungs with the sweet resinous odour of the pines. Everywhere wild irises were growing, and behind the rest-house the hillside was one glorious splash of blue and purple. The place deserved better acquaintance than I was able to develop during my brief halt, and I resolved to come back and spend a few weeks in this haven of peace and beauty.

The jeep was ready and we started again. The road lay through a lovely thick forest of fir, Himalayan oak and rhododendron. Irises grew everywhere, interspersed with blue and white anemones. A mile from Shoja the jeep stopped again. I stepped out and began to walk, telling the driver to follow as soon as he was ready. This manœuvre was repeated several times. I was glad to have an opportunity of exercising my leg muscles and keeping them in trim for the trek, though I had scarcely anticipated such an ample measure of pre-liminary conditioning. I must have walked a total of more than ten miles during the several stoppings of the jeep. As the day advanced, the route became hot and dusty, for we were down to an altitude of less than 4000 feet. The last time we had engine trouble I walked nearly six miles before the jeep picked me up. I was beginning to think that I should have to walk all the way to Banjar, perhaps to Out, where I should arrive at midnight, and then go on to Kulu. I wondered how I could send a message to my friend Shrinagesh, the Commissioner of the Division, or to Bachittar Singh, the Sub-Divisional Officer. There was no vehicular traffic and the only human beings I met were some men

of the Public Works Department working on the road.
They greeted me with a most aggressive *Jai Hind*, but
they could not be of any assistance to me. It was,
however, gratifying to note that even in these back-
woods our people had become conscious of their newly
acquired freedom in less than three years, and had
adopted *Jai Hind* as the form of universal greeting.
They thus paid homage to their motherland each time
they met a fellow Indian. We, from the towns, alas !
had not yet changed our ways and had not given much
heed to the government's directive to use *Jai Hind* as the
official form of greeting.

Afterwards I learnt that the road coolies and villagers
had all been carefully and assiduously drilled to say *Jai
Hind*, instead of using the older form of salutation.
Regimentation has its points : the residents of Kulu
continue to say *Jai Hind* to each other and derive a
certain measure of emotional satisfaction in remember-
ing their country whenever they meet a friend.

A yellow speck appeared round a bend in the road
and in a few minutes the jeep was carrying me towards
Banjar. The driver confided to me that the platinum
points had gone completely, and it was only his extraordinary skill which had enabled the engine to start and
to continue working. I was greatly impressed and
registered admiration.

We arrived at Banjar at a quarter past two. The
village is large enough to deserve being called a small
town. It is the headquarters of a sub-*tahsil* and a police-
station. It has a civil dispensary and a veterinary hospital and a long narrow bazaar with over a hundred
shops. In the shops and in front of them lay quantities

of *koot*,[1] salt, raw wool, *gur*,[2] brightly coloured cotton fabrics, cakes made from tobacco, kneaded with molasses to make it fit for the *hookah*, stale sweets and *pakauras*.[3] And over and around everything were flies — millions of them — humming a continuous nasal note and forming a quivering network across the sunlit air. They flocked round the jeep and alighted on my hands and face.

I wished I had insisted on going to the rest-house, instead of letting the driver stop in the middle of this filthy bazaar. But since we were stopping for 'only two minutes, *sahib*' I had raised no objection. The two minutes lengthened into half an hour. Somebody brought a chair and a folding-table which refused to unfold. Would I like to eat something — a *jalebi*[4] or perhaps a *samosa*?[5] I thought of all the flies which must have deposited their excreta on every *jalebi* and every *samosa* in the place, and tried to contort my face into the semblance of a smile while I declined the kind offer. There was a spring of fresh water issuing from the hillside straight into the bazaar and I felt safe in asking for a drink. A glass tumbler was brought. I washed it in the spring for a whole minute and took a long cooling draught. I wanted to eat a sandwich ; but there were so many spectators standing and staring at me that I felt like a poor actor who is frightened of saying his part because he knows he will cut a sorry figure.

[1] Aromatic root of a Himalayan bush used in the manufacture of incense. [2] Raw sugar.

[3] Chopped vegetables dipped in batter and fried in mustard oil.

[4] A kind of sweets, crisp and syrupy, delicious when fresh but wholly unpalatable when stale. [5] Savoury patty.

A tall man in a shabby suit, several sizes too small for him, came up and gave me a most hearty *Jai Hind*. He was obviously some official, not smart and self-possessed enough to be a head-constable of police ; perhaps a road-, or excise-inspector, or someone from the *tahsildar*'s court. He wanted recognition. I acknowledged his greeting and then ignored him completely. He stood in front of me for a few moments and then moved away, pulling awkwardly at his finger-nails and twisting his neck from side to side to release it from the bondage of his shirt collar. I have no doubt he cursed me for an arrogant judge who had not a kind word for a humble servant of the Republic. But I was hot and hungry. I had sent several messages to the driver and the *tahsildar*, and the long wait had drained all the milk of human kindness out of me. Perhaps one day I shall be able to make amends to the disappointed official, and speak kindly to him.

The driver and the *tahsildar* finally returned, wiping their mouths and apologizing profusely. We left Banjar at three. As we drove out of the bazaar, the Tirthan stream suddenly came into view. Perched above an eminence on the right bank stood the P.W.D. rest-house, ideally situated to command a view of the entire valley. Across the Tirthan the hill rose steeply for several thousand feet, and almost at the very top lay a beautiful village of flat-roofed houses and white-washed walls. The distant pattern of tiny white squares on a background of grey and green made a very pretty picture.

We descended by a series of hairpin bends to the bridge over the Tirthan and sped onwards to Larji at a good pace. The road henceforth was broader and straighter. We had to stop once at a bridge which, the P.W.D. notice informed us, was not strong enough to bear the strain of a loaded vehicle running over it. So we walked across while the jeep was driven over at a slow speed.

At Out there is a good strong bridge over the Beas, and then the road runs for some miles through a narrow and picturesque gorge. The valley opens out near Bajaura. Here I stopped to see the temple of Basheshar Mahadev — a beautiful monument only a few minutes' walk from the road. The shrine is dedicated to Shiva and dates probably from the eleventh century A.D. It is constructed entirely of stone and has exquisitely carved slabs more than five feet in height, set in projecting porches, on the north, west and south, the doorway being on the eastern side. The bas-relief of Vishnu in the western porch and the figure of Durga, represented as slaying two Asura kings and the buffalo demon, on the northern side, are carvings of great beauty and excellence. The tall figures are chiselled out with perfect grace and the detail is finished with consummate skill. Inside the sanctum is a Shivling. The capital was damaged by the earthquake of 1905, but it has been replaced by a new stone carved exactly like the old one.

A slight drizzle was beginning and I hurried back to the jeep. We rushed along through villages and fruit orchards. At Buntar the *tahsildar* pointed out the landing strip near the river-bank, prepared last year at the instance of Shrinagesh. It was barely visible through the gradually thickening veil of drizzle. By the time we reached Kulu it was raining steadily. As the jeep stopped in front of Calvert Lodge, the residence of the Sub-Divisional Officer, one of the tyres gave a loud hiss and collapsed. The *tahsildar* reminded me that my exalted office had acted as a talisman during our journey.

It was undoubtedly an achievement to reach Kulu in a little over twenty-four hours. The route I had followed involved a journey of only 120 miles as against nearly 450 miles by the usual way via Ambala, Jullundur, Amritsar, Pathankot and Mandi. It had meant a saving of a whole day, and I was spared the fatigue of a very tiresome journey. It had also meant a few thrills and moments of apprehension. But these had served to provide a certain measure of interest, so that I could say : *Je ne suis pas allé, j'ai voyagé.*

II

IMPEDIMENTA
AND IMPEDIMENTS

I WAS the first to arrive.

Bachittar Singh, the Sub-Divisional Officer of Kulu, welcomed me and said that Shrinagesh and the Bhavnanis were due at six o'clock.

They arrived at eight o'clock.

Our party was now complete : Shrinagesh, Mohan Bhavnani, Chief Producer of Documentary Films for the Government of India, his wife, Enakshi, and their son Ashok, Bachittar Singh and his son, Pal, and myself. We all had our parts assigned to us. Shrinagesh was to lend his authority as Commissioner, on official duty ; Bachittar Singh as Sub-Divisional Officer and Officer-in-Charge of Settlement Operations [1] was to look after the general arrangements and act as liaison officer ; Enakshi Bhavnani was to control and issue provisions and organize the food side (she had brought up vast quantities of tinned stuff from Bombay, where the stocks are fresher and cheaper) ; Bhavnani was to take still photographs and a coloured movie film (each of us had our own camera also, and we were to exchange photographs at the end of the trek, so that everyone should have a complete photographic record) ; I was

[1] See footnote on p. 102.

entrusted with the task of supervising the setting up of the camp each evening, and seeing that the tents were dismantled, packed and sent off in the morning without delay. This was particularly important as, on a trek of this kind, one must make an early start each morning. After three days our men acquired so much skill and dexterity in putting up and taking down the tents that like the two boys, Ashok and Pal, I found myself jobless, and gave myself up completely to the joys and pleasures of the trek.

The morning of July 1 was bright and warm. I went through my yoga exercises and had the satisfaction of feeling that my lungs, heart and muscles were in good trim. During the days that followed I realized that I had the advantage of coming from Simla, which is 7000 feet above sea-level, and, in consequence, felt the strain of going up to greater heights much less than the others, who had come from the plains and had not been leading such an active life as a person living in the hills. Also the practice of *pranayama* (yogic breath control) had increased my lung capacity and I could climb fairly fast without getting winded. Nevertheless there were moments during the trek when I regretted not having learnt a little more of the science of yoga and acquired a little more skill in the art of controlling the bodily organs. Even a small fraction of what the *tantrics* know and practise would have been useful when mind and body felt limp towards the end of a long and exhausting march, and when the few hundred yards to the journey's end seemed an insurmountable barrier between oneself and one's greatest desire. How often I wished I had the power to hitch my spirit to a distant star and walk fifty miles over and across mountains without feeling any kind of fatigue, or had been able to generate so much internal heat that a wet sheet wrapped round the bare body would dry in a short time, though the atmospheric temperature might be below freezing point. But since I did not possess these powers, I could only toil along and let my instinct for self-preservation do what yoga might have helped me to achieve with far greater ease.

After breakfast I went over to check the tents. We had six double tents lent by the Survey of India Department. These were very compact and weighed 40 lbs. each. We also had fourteen *chholdaris.*[1] These were single tents but very roomy and very comfortable. I asked the men to put up one of the double tents. It was a brief and simple process. The two uprights and the cross-pole were each in two pieces. These were placed end to end and fastened by an iron ring which slipped over the join. The cross-pole was placed in position between the inner and outer canvases, and the pegs of the uprights were coaxed into eyelets at each end of the cross-bar. The tent now stood up with its flaps hanging loose. As soon as the ropes were pulled taut and tied to pegs which were hammered into the ground with a heavy wooden mallet, the tent was ready. It was large enough for two small beds laid side by side, or you could have a bed on one side, and a small table and chair on the other. Two persons could live inside it in comfort, and one in absolute luxury.

Each tent had a waterproof ground-sheet and a

---

1 *Chholdari* — a single tent generally used by servants and coolies.

striped cotton carpet to spread over it. It took three men fifteen minutes to put the tent up. For six tents and fourteen *chholdaris* we should need more than two hours. I had the tent taken down and packed up. Then I asked four men to unpack it and put it up again as quickly as they could, while I kept looking at my watch in a determined manner. This time the tent was up in less than nine minutes. This was much more satisfactory. During the trek we were able to reduce the time still further, and whenever we camped near a village and local help was available, the whole camp would be up and ready in three-quarters of an hour. We never had any trouble in breaking up camp once we made it clear that the mules must leave before we did. The mulemen could have either stayed behind, cooked their food and taken it with them, or reached the destination early and done the cooking there. We ruled out the first alternative ; thus we were able to make an early start each morning.

There was a whole heap of furniture, camp-beds, folding-chairs and tables, lanterns, Petromax lamps, a wireless receiving-set, a wireless transmitter complete with batteries and two dynamos to charge up the batteries, petrol-cans for the dynamo engines, two four-gallon canisters containing paraffin, a basket of fresh eggs, a large basket full of live chickens, crates full of tinned provisions, canisters of clarified butter, sacks of flour, charcoal, potatoes, onions and other vegetables, a crate of apples, trunks, boxes, bedding-rolls. It was like the impedimenta of an army preparing to move. I was appalled at the quantity and unwieldiness of our luggage as it lay sprawling over a large area of the grassy field in front of the S.D.O.'s office. We should require a hundred pack-mules and as many coolies to transport all of it.

I asked Bachittar Singh what was the total strength of our party, including servants and others. He began to count :

'There are seven of us. Then there is Mr. Shrinagesh's H.V.C. (Head Vernacular Clerk) and his stenographer ; then there is my reader, and two orderlies ; the cook, Mr. Shrinagesh's bearer, his orderly. How many is that ? Fifteen. The doctor is joining us at Manali, and of course there's the wireless operator and finally the muleteers. I don't quite know how many of them there will be.'

'The doctor ?' I asked. 'I suppose we shall need one with us.'

'He is only going with us part of the way. You see, every other year, a Sub-Assistant Surgeon is posted in Spiti for three months during the summer season. Last year no one went, this year Dr. Massey has been chosen for the post, or rather he has chosen himself. The head of the medical department asked for volunteers, and Dr. Massey, being a good Christian, offered his services. It is convenient for him to go with us, instead of making his own arrangements separately, and, of course, it is convenient for us. He is waiting at Manali with three boxes full of medicines.'

I said : 'I see ; more luggage. How many mules shall we require ?'

'I have arranged for forty-six and they should be enough.'

'Ah !'

c

That was all I could say.

In the end we had fifty pack-mules and six riding-ponies. The slow-moving train of our caravan was a heartening sight in the wildernesses of Spiti and Lahoul. It was reassuring to feel that one was not alone while facing the cruel and terrifying grandeur of the Himalayas.

It was eleven o'clock before I had finished checking the tent equipment, and I had to abandon a tentative plan to visit the shrine of Bijli Mahadev. I had been intrigued by the legend I had heard, and wanted to see the Shivling which is split into two every year by a stroke of lightning, and is then miraculously joined together by the priest of the temple. The lightning, representing the blessing of heaven, is released from the trident of Lord Shiva, and enters the temple through a tall staff, sixty feet in height, made from a single fir or spruce trunk, which stands in front of the temple. I could see the staff shining like a silver needle growing on the top of the hill across the Beas. Going to the temple involved a walk of eleven miles and a climb of 4000 feet. If I had started early I might have been able to do the twenty-two miles of the complete journey and been home for a late tea. Even so it would have been a hot and tiring march in the sultry atmosphere of July, and I was not sorry that the tents prevented me from paying a visit to Bijli Mahadev.

Like all earnest people, Shrinagesh has a hobby-horse and in his case it is air travel. The greatest obstacle to the development of the Kulu valley is lack of good and fast means of communication. The direct road from Simla by which I had come is, at the moment, fit for jeeps only. The Mandi-Pathankot route is long and commercially unprofitable for the transport of goods. The portion of the road between Mandi and Out is frequently closed by landslides during the monsoons, so in the months of July and August when fruit is ripe and ready for export, there is often no means of sending it out. Every year about a third of the total fruit crop of the valley rots and goes waste in this manner. Visitors who wish to come to Kulu for a holiday keep away when they hear stories of the valley becoming isolated for weeks at a time. Those who come in early summer are in a hurry to leave after the first monsoon shower. These fears of being cut off from outside contact are sometimes exaggerated, but they are certainly not groundless, and means of communication with the valley are far from satisfactory.

This is very regrettable, because Kulu deserves to be known and appreciated better. The climate of the valley is salubrious without being either relaxing or what is called invigorating, which usually means uncomfortably cold. The soil is so rich that vegetables and flowers of every variety flourish and grow to enormous sizes. There is fruit in abundance, and of the very best quality. Apples, pears, plums, apricots, cherries, strawberries, walnuts and chestnuts are grown with excellent results. There is almost unlimited scope for fruit-canning and jam-making industries. There are hot springs, possessing medicinal properties ; at Manikaran they are said to be radioactive, and at Bashisht they have a wholesome sulphur content. The whole valley is full of beauty spots and is an ideal holiday resort. But because it is difficult of access it is not popular and only a few visitors venture to go there and taste of its sweets. Even more regrettable is the loss of national wealth when fruit weighing thousands of tons rots on the trees and perishes because of road breaches.

These considerations gave Shrinagesh the idea of connecting Kulu with the rest of the Punjab by a regular air service. During the war a small plane had succeeded in landing on a natural strip of flat grassy plain on the right bank of the river Beas near Buntar. The strip could be enlarged and improved by clearing it of boulders which lay strewn all over it. If it were possible to land a Dakota or other aircraft of similar size at Buntar, the valley would remain open despite landslides and road breaches, and the fruit-growers could send out their fruit at a small additional cost. So Shrinagesh had the strip extended to nearly a thousand yards with funds contributed partly by the District Board and partly by the fruit-growers.

Near Gondla also, in the Lahoul valley, across the Rohtang pass, there is a natural strip lying deep down in the valley, on the right bank of the River Chandra. This strip is nearly a mile long and more than a hundred yards broad. Shrinagesh had seen it the previous year and had put men on to clear it of stones and boulders. If a plane could land at Gondla, Lahoul would be open to visitors, and even comfort-loving tourists of the arm-chair type would be able to see this glorious valley. But the Civil Aviation authorities were doubtful. They feared that the Kulu valley at Buntar was too narrow and the hills enclosing it too high to permit access by large aircraft with safety. Shrinagesh had travelled to Kulu several times by air, but he had always come in a

Beechcraft Bonanza. He wanted to fortify his argument with the opinion of a recognized expert, and had asked Mehr Singh — India's best pilot — to come over to Kulu and give his opinion and advice regarding the capabilities of the Buntar air-strip. He might also be able to have a look at Gondla, and tell us if it were possible to land a plane there.

Mehr Singh was expected at four o'clock. This was the time Shrinagesh had mentioned in a wire which he had sent early that morning. At twelve o'clock we heard the drone of an aeroplane engine, and saw a tiny speck of silver circling in the sky towards the south. A jeep was sent off to Buntar at once and in less than half an hour it brought Mehr Singh to Calvert Lodge, fat as ever and laughing merrily. He wriggled out of the jeep and asked Shrinagesh why he hadn't sent him the promised telegram.

I had heard a great deal about Mehr Singh and I was delighted to meet him and make his acquaintance. I found him a cheery person full of fun and *bonhomie*. For a man whose skill and courage as a pilot has been so much talked about and advertised, he was remarkably modest, and in his conversation seldom referred to himself. He had a large stock of stories, including some very naughty ones, and we spent a most hilarious evening, going into continuous fits of helpless laughter.

After tea we drove down to Buntar to have a look at the plane. It was a lovely thing — a two-seater Texan — all shiny and trim as it stood on the grassy down, surrounded by a large crowd of admirers from the village of Buntar. A police constable was pushing them back to a safe distance, but it was all he could do to keep them from rushing towards it and scrambling over it like a swarm of bees.

Mehr Singh took Shrinagesh up to try and have a look at Gondla. But low clouds hung over the valley to the north and they could not go even as far as Rohtang. The plane returned after twenty minutes and came down skimming the tree-tops, fifty yards from the edge of the airfield.

'You see what I mean,' he said as he scrambled out, 'you must have those trees cut; otherwise the plane can approach from one side only.'

Mehr Singh thought the strip was long enough for aircraft of the Dakota class, but the valley was so narrow that it would be inadvisable to bring in a big plane. There wasn't room enough to take a sufficiently wide circle when landing or taking off, he said. In favourable conditions a Dakota would be perfectly safe.

'I suppose any pilot would be able to bring a Dakota in here?'

'Any good pilot,' said Mehr Singh.

We laughed indulgently. Mehr Singh's standards were perhaps too high. With his daring and his almost uncanny skill in handling aircraft he had done things which most other pilots would consider impossible. His landings in Poonch and in Leh during the Kashmir operations will always be remembered with awe and admiration. Did he think a 'good pilot' meant a pilot who was as good as himself?

Shrinagesh was disappointed. The Civil Aviation authorities were right, and there was no prospect of his air-strip being certified fit for commercial air services.

Our luggage and tents had been sent to Manali by

lorry in the course of the afternoon, and we had an evening of untroubled ease, anticipating the joys of the trek and listening to Mehr Singh's stories. After an early dinner we drove the twenty-three miles to Manali along a bumpy road and arrived there at half-past ten. It was delightfully cool in the Forest Department rest-house and our beds were ready to receive us. I went to sleep at once.

The next morning was bright and clear, and as I sat on the verandah sipping my cup of morning tea, life seemed good and full of promise. There was, however, trouble ahead, and it was not long before we felt the impact of mischance, inexperience and incompetence.

We wanted about fifty pack-mules and six riding-ponies. The mule contractor told us that he had been asked to supply only forty-six mules (there was no difficulty about riding-ponies, we could have had a dozen if we wanted them). But there were only twenty-five mules available in Manali. The *tahsildar* looked foolish and helpless. He ran from the Forest rest-house to the P.W.D. bungalow in the bazaar and back again several times, but his exertions bore no fruit. At ten o'clock, the hour we had planned to start, there were just twenty-five mules and no more.

The muleteers raised an objection to the size and weight of the packages. We should have to travel along steep and narrow paths, and cross several high passes. Pack-animals could not carry such heavy loads, and in some places the path was so narrow and ran so close to the hillside that a bulky load would foul the rocks and endanger the life of the animal.

As we were considering this matter, an orderly came

and told Shrinagesh that one of the wireless operators was lying seriously ill. Dr. Massey was sent to examine him. He returned in a few minutes and reported that the wireless operator was perfectly well, and that there was nothing whatever the matter with him. Someone now informed us that the man had a young wife in Palampur, and he had been scared by stories of hardships and privation in Spiti and Lahoul which the muleteers had been relating to everyone. Shrinagesh and I went across to the garage where the sick man was lying in bed. On seeing us he set up a full-throated concerto of frenzied groans and howls of such vigour and amplitude that we could not but admire the excellence of his performance. He complained of aches in all parts of his body and said it would be quite impossible for him to undertake such a long and arduous journey. He said his companion, Joseph, the other operator, was willing to go alone, and he was quite capable of looking after and handling the transmitter single-handed. Joseph, who had no wife in Palampur or indeed anywhere else, agreed with the sentiments expressed by his colleague, and so we left the matter at that.

The problem of loads, too, was soon solved. We procured pinewood planks and the services of a carpenter. He set to work, and in less than two hours he had nailed together enough crates of the required size to carry all our stores and provisions. The original boxes were abandoned and their contents were transferred to the newly constructed crates.

News was brought that a party of mules was on the way, and was expected to arrive within an hour or so. Things began to look hopeful, and at half-past ten we drove over to Major Banon's bungalow to have our breakfast.

Major Banon is undoubtedly the most important, and perhaps the most interesting, person living in the valley. He owns an extensive orchard which was planted by his father, Captain Banon, in 1884. It is said that Captain Banon came to Kulu on reading an account of the valley which moved him so deeply that he travelled all the way from Bengal to see this beauty spot. He fell in love with the place, and decided to settle down in Manali. He took a beautiful hill woman as his wife, and began to cultivate a variety of excellent fruits. His three sons succeeded to his property. All of them are married to Kulu women, each having more than one wife. The eldest is Major Banon, whom the local people call Chini Sahib; his two brothers are known as Kathaura and Mishter. All three speak the local dialect like natives; indeed, Kathaura and Mishter experience some difficulty in expressing themselves in English. They are all Indian citizens and their declared religion is Hinduism. They know that Hinduism permits polygamy.

Major Banon is the *sarpanch* [1] of the local *panchayat* [2] and the president of the local co-operative bank. He enjoys the confidence of the people, and wields considerable influence over them. They look upon him as a sort of super Kuluite who, by his education, knowledge of English and the white blood in his veins, can solve their problems for them and act as intermediary between them and the government. During the British

---

[1] Chairman.
[2] Elected body of village elders, exercising administrative and judicial functions.

regime his position was even more important than now.
There can be no doubt that he has taken root in Kulu
soil, and though he runs a fairly profitable business as
fruit-grower he often allows himself to be swindled by
unscrupulous servants. He takes in paying guests, and
during the summer months he has as many as twenty
people at a time staying with him.

His knowledge of local folklore is both extensive and
deep. He told us of the goddess, Hirma Devi of Manali,
on whose caprice depends the celebration of the annual
Dussehra festival at Kulu. Although it is Raghunathji
who presides over the Dussehra festivities, the proceedings cannot start until Hirma Devi condescends to leave
her abode in Manali and travel down to Kulu. Indeed
it is her presence which determines the entire course of
the ceremonies.

We asked : 'And how does anyone know if she is
going to be perverse and stay in Manali, or go down to
Kulu and let everyone have a lot of fun ?'

'She speaks through her priest, who goes into a
trance and acts as her oracle.'

I was once present at a séance of this type when a
priest was communicating the wishes of his *deota*.[1] I
was amazed and wonder-struck by the performance of
the possessed man. The *deota* was borne on a palanquin,
carried on the shoulders of two men ; the priest stood
near the palanquin, his hand resting lightly on the hem
of the cloth which draped the *deota* like a long skirt. The
palanquin began to rock gently from side to side. This
was a sign indicating that the *deota* was ready to speak.
He had perhaps something to say, or was willing to

[1] Deity, minor god.

answer questions put to him.   Suddenly the priest began
to tremble all over, his face assumed a look of pain, and
large beads of sweat stood out on his forehead.   The
trembling became more violent, and incoherent sounds
began to issue from his drawn and rigid lips : 'Ta, ta,
ta, ga, ha.'   Then came phrases and sentences in a rapid
flow of intelligible language : 'The rains will be delayed, but when they come there will be an unusually
heavy downpour.   The barley crop will be average to
good'.

People crowded round the priest asking him all
kinds of questions.   'When will my son get well ?'
'My cow doesn't calve ; what shall I do ?'   'Beli won't
pay me my fifty rupees, shall I bring a suit against him ?'
'When should I buy the wool for my shawls ?'   To
some there was a reply, others were ignored.   A man
pushed his way forward and asked : 'What is the most
propitious time to build my house ?   Every time I lay
the roof it falls down.'   The priest continued his con-
vulsions and hardly looked at the man, but the reply of
the *deota* came back clear and unmistakable :

'You will be able to build your house when you
bring back the pound of nails which your father stole
from the temple.'

I did not doubt that Hirma Devi, too, spoke to her
devotees through her priest in the same way.

There was just enough time to pay a brief visit to
her shrine.   She lodges in a pagoda-shaped temple with
a three-tiered roof in a gloomy deodar forest, just beyond
Major Banon's house.   The deodars are the tallest and
the most magnificent I have ever seen.   The trees must
be over a thousand years old.   Some of them were cut

down and removed for timber a few years ago, but a somewhat eccentric Scottish member of the Indian Civil Service fortunately stopped further depredation by issuing an order that the deodars were 'ancient monuments' and therefore protected under the Ancient and Historical Monuments and Archaeological Sites and Remains (Declaration of National Importance) Act. No one challenged the legality of the order and the trees were saved.

The temple doorway is elaborately carved. The interior is dark and forbidding. Through the half-open door one can see a number of large boulders on the floor, and a huge stone slab which seems suspended in air. A rope hangs from the roof. Tradition has it that in olden days human sacrifices were performed in the temple, and the victims were suspended by this rope and swung over the head of the goddess. Now the temple is sometimes used for punishing a refractory deota, in times of drought, if prayers and incantations fail to persuade him to do his duty. The deota is locked up in the dark cell till he comes to his senses and gives to the people of the valley their due measure of rain.

At Major Banon's we met Horace Alexander, who told me that the Oxford University Press had asked him to write a book on the Rehabilitation of Refugees in India, that he had just finished reading my Stern Reckoning — in fact, he informed me with a backward jerk of his head, the Prime Minister's personal copy was still in his possession; he had been for a trek in Lahoul and had seen all the goats and sheep dying of foot-and-mouth disease; really, something ought to be done about it. I was duly impressed and agreed with him.

We also met a tall young Englishwoman with badly
blistered lips. She, too, had been trekking in Lahoul.
The afternoon winds, she said, were something dreadful.
They blew with such fierce intensity that dust particles
flew up and hit you with the force of shots fired from a
gun — witness the bad blisters on her lips. She gave
me a Survey of India map of Spiti and Lahoul, which I
badly needed. I had not been able to get one in Simla
and was glad to have it.

After a large and very sustaining breakfast of soup,
rice, meat-ball curry, egg curry, several vegetables, *chappatis*,[1] both baked and fried, fruit-salad and cream, and
fresh fruit, we drove back to the Forest rest-house. We
were greatly relieved to find the rest-house empty. All
our luggage had been taken down to the *dak* bungalow
where it was being loaded on mules.

At last, we thought, we had enough mules, and could
make a start. This was good news and we hurried to
the *dak* bungalow. The gate of the rest-house was shut,
and Ashok Bhavnani jumped out of the car to open it.
He stepped into a hole and twisted his ankle. As he sat
down, helpless with pain, there was an anxious cry from
his mother. Ashok was lifted back into the car, and
taken to the *dak* bungalow. Dr. Massey's services were
once again requisitioned, and Ashok's foot was wrapped
up in masses of cotton wool and bandaged round and
round till it grew to a fabulous size.

Ashok was upset because he would not be able to
walk for a few days, but this was not a matter about
which any of us was greatly worried, because a riding-
pony would be able to carry him just as far and just as

1 Unleavened bread, flat and round like a pancake.

fast as his own feet ; what we *were* worried about was
the supply of mules. Our informant at the Forest rest-
house had been more hopeful than accurate in what he
had told us. There were only forty mules at the *dak*
bungalow. These had been loaded and were ready to
start ; but there was still a lot of stuff lying scattered
about. The contractor swore that a large herd of mules
(it would be more appropriate to say an obstinacy or
perhaps a procrastination of mules) was on the way.
They had crossed the Hamta the previous day and
should be arriving in Manali any moment.

We waited impatiently till two o'clock, and then
decided to send off the mules which were ready. This
would greatly ease the situation by reducing confusion
and putting an end to the loud and abusive arguments
with which the muleteers were expounding the merits
and demerits of each other's animals. It was also decided
that some of us should start, as we were doing nothing
useful by sitting about and fretting, and the best way
to cure frayed nerves is to force oneself into some form
of physical activity.

So Shrinagesh and Bachittar Singh stayed behind to
complete transport arrangements, while the rest of us
set off on the first lap. The Bhavnanis and Pal mounted
their ponies, and I took to my feet. We had to travel
only nine miles, and expected to arrive at the Chikka
camping-ground at seven or thereabouts. We should
have time to choose a good site for our camp, have the
tents put up, and the dinner cooked. By then Shrinagesh and Bachittar Singh would arrive with the remaining luggage. It seemed an excellent arrangement and
very simple.

III

THE HAMTA PASS

IT seemed simple, but as I went down the steep path to the picturesque cantilever bridge of pine logs spanning the Beas, I asked myself if our decision were wise, and wondered just how it would work out. What would happen if no more mules came? We had not even envisaged this possibility, otherwise we might have re-arranged the loads, and discarded anything that was not absolutely essential. Instead, we had sent off four-fifths of our luggage, including articles that we might well have done without, while the remaining one-fifth comprised a wireless transmitter which Shrinagesh considered most essential, and a number of packages which (I learnt before nightfall) contained poles and tent-pegs without which the tents could not be put up. I tried to allay my fears by telling myself that the Commissioner of the Division and the Sub-Divisional Officer, working to-gether, made a formidable team, and nothing was beyond their powers. They would certainly be able to do some-thing, and *faute de mieux* they would press into service a gang of coolies. In any case *I* could do nothing about it, so I tried to console myself with the thought that there were so many important officials concerned in this venture that everything must have a satisfactory conclusion, as the *tahsildar* and the driver of the jeep had so often told me.

33

After crossing the Beas, the path turned sharply to the right, and for some distance proceeded along the left bank of the river. All along the way lay well-watered paddy fields in which men and women were busy planting rice seedlings. Each worker carried a sheaf of tender green shoots and, wading through ankle-deep mud, moved along his or her row and pushed the seedlings in, one by one. It must have been cold and tiring work, squelching through pools of mud and remaining bent double for hours together. Every now and then a worker would straighten himself and press in the small of his back with his free hand. I passed through Parini, a neat and compact village, as, indeed, are all villages in the Kulu valley, and turned to the left — a villager told me that our mules had gone *that* way. The path skirted a hill that rose steeply to several thousand feet, and began to climb sharply up the face of a rugged mountain. I went up and up, scrambling over stones and boulders. The mules and horses were toiling up the gradient in a long slow-moving procession, like a serpent crawling up the hillside. Every few minutes the whole train of animals would stop to take rest. The muleteers would stop too, and stand motionless, as if petrified by a strange spell. Suddenly one of them would spring into activity, utter a loud yell, and hit the nearest mule hard with his stick. On this the entire line would become animated once again, and move forward. A hundred yards higher up there would be another halt, and the ritual would be repeated.

I passed the Bhavnanis, and asked Ashok about his ankle. He said it was much better, but complained that his foot had been padded up to such a size that he could not put it in the stirrup-iron. I found I was able to maintain a fairly good pace, and passing the mules, went ahead and overtook our cook, Chanchlu, who had been sent off from Manali at an early hour, so that he should arrive in time to prepare the evening meal.

Chanchlu was an extraordinary character. He was what his name implied — the restless one. In appearance he was one of the unloveliest creatures I have ever met or seen. He was short-statured, with small spindles for legs that remained perpetually flexed at the knees. I never saw him straighten them fully whether he walked, stood, sat, rode a pony, or lay down to rest. This made him look even more diminutive than he really was, and, at the same time, invested his movements with a comic and risible aspect. His skin was very dark, almost black, and of a dull smoky texture. His features were cast in a simian mould, and when he showed his teeth the splash of white appearing suddenly out of a dark nothingness at first shocked you, then it made you wonder, and finally compelled you to laugh. But Chanchlu was a cheerful soul, and I never heard him complain. He was up early and had the fire going at four o'clock each morning ; I do not know at what time he went to bed, for I used to hear him pottering about in the kitchen-tent long after we had retired, and sometimes, if I woke up in the middle of the night, I heard the clinking of pots or the swishing sound of pans being scoured clean with ashes from the kitchen fire. I do not think he slept more than two or three hours a day. Yet he was always active and was certainly the hardest-worked individual of our party.

His cooking was excellent, though it lacked variety.

D

He could turn out an omelette from dehydrated egg-powder which looked and tasted as nearly like a dish of fresh eggs as I ever hope to sample. His manner of dealing with sausages was entirely above reproach and when, one evening, he served us with what he called 'tasty mutton steak' we thought we had attained the acme of gastronomic ecstasy. But his *forte* was a sort of all-embracing, inclusive omnibus dish which was called hotch-potch. It was soup, entrée and joint all combined in one. Its ingredients, as far as we could fathom, were water, rice, some kind of pulse, a few pieces of meat, more water, potatoes, peas, *baris*,[1] spices and more water. It arrived on the table in a large china bowl and we ladled out large helpings of it into our soup plates. This we consumed with great relish, taking repeated helpings till we were satiated. Hotch-potch made its appearance on our dinner-table with a frequency that bordered on prodigality. There was always plenty of it, for the supply of water in a region of snows and glaciers is constant and unlimited. Towards the end of the trek we began to look upon the very name of hotch-potch as a kind of bogy with which we could subdue and vanquish the hydra-headed monsters of epicurism and gluttony.

Chanchlu was a most resourceful person, and never at a loss for anything. If he was tired he selected a lightly laden mule and perched himself above the load. He paid no heed to the protests of the muleteers, and pretended he hadn't heard what they were saying. His office as cook, no doubt, helped him in stifling much of

[1] Dehydrated balls of a highly spiced preparation of which the chief ingredients are mashed pumpkin and ground pulses.

the criticism, for there was always something in the pot which could be smuggled out to anyone who had won his favour, or whose favour he wanted to win. He was not impervious to Cupid's darts, and never wasted an opportunity to start a mild intrigue or chase the contents of a willing petticoat. Altogether Chanchlu was a very human and likeable person.

I came upon him now, on this first day of our march, plodding slowly along his way, and working his thin black legs like an ant carrying a burden many times its own weight. He offered to carry my wind jacket and cameras, but I told him I was bigger and stronger than he was. He laughed and said he could walk as fast and as far as any hillman. He had not gone on ahead as he was worried about the *sahibs* being so late.

The steep climb ended and we came to a grassy down, strewn with masses of blue iris and kingcups. Wild strawberries grew all over the place. I picked a handful and ate them. They were delicious. There was a spring of ice-cold water issuing from underneath a stone, and splashing into a tiny pool below. I drank deeply of it, and rested for a few minutes.

A man brought a message from the Bhavnanis that they were having tea and were expecting me to join them. I asked how far behind they were.

'About a mile,' said the man.

I was hungry and could have done with some tea and sandwiches which I knew were in the tea-hamper, but I had not the stomach to walk back a mile and toil up the steep hill once again, so I ate a few more strawberries, had another drink of water, and set off in the company of Chanchlu.

We passed through a fir wood and came to a clearing where the ground was littered with trunks of fir-trees, lying dead and battered, as if some monstrous giant had plucked them up by the root, sucked them dry of their sap, and thrown them on a rubbish dump. An avalanche of snow, descending from the steep hill on the right, had brought them down and left them to rot at the bottom. The mass of snow had come hurtling down the steep slope, and crashed into the stream at the bottom of the valley with such tremendous force that the resulting blast had torn up the trees on the opposite hill, up to a height of more than a hundred feet. There could be no other explanation of the broad grassy lane which ran down the hill from one side, continued its way across the narrow valley and up the hill beyond, for a hundred feet, and then ended abruptly. On either side of this lane grew a thick forest of fir and cedar. Numberless goats and sheep were grazing everywhere, and the ground was covered thick with their droppings. The *gaddis* (goatherds and shepherds) were camping in the lee of a huge rock and cooking their food. The sound of their palms slapping the dough, and flattening it out into large round *chappatis*, before turning them over to bake on a concave iron girdle, was pleasing to the ears. A little farther we met a man carrying a four-gallon canister on his head. He said he had been waiting with our milk supply since ten in the morning, and was now returning home. He was at first reluctant to turn back and accompany me as he had a long way to travel, and it was getting late ; but I succeeded in prevailing upon him by making promises of food and shelter for the night.

The path continued up a gentle incline, hugging the Hamta torrent, which went roaring down the narrow valley on our left. At one place the valley was completely bridged over with snow frozen hard to a depth of several feet. Here the Hamta had carved out a huge dark cave with walls of ice ten feet thick into which it plunged headlong with a loud resounding groan. A fine spray played over the mouth of the cave, and hung in the air like sparkling mist.

Near this fairy scene a party of *gaddis* had lit a fire, and were preparing to cook their evening meal. A brass *degchi*, scoured and washed till it shone like a mirror, had been made ready to cook a dish of potatoes. I borrowed this and filled it with milk from the canister. This I had boiled. As I drank it slowly out of a brass tumbler I felt warmth and energy pouring into me, and with every sip the marvellous cave of ice became more enchanting ; the continuous groaning of the Hamta torrent sounded like the drone of a *tanpura* [1] under the soft touch of a master musician.

But it was getting late, and I had to see to the tents. The milk had revived me completely and in half an hour I arrived at the place where our mules had stopped for the night and unloaded. The Chikka camping-ground was still another mile away, but to reach it we should have to cross the Hamta stream by a narrow pine-wood bridge scarcely two feet wide and unprotected by side railings. The muleteers said it would be extremely risky for pack-animals to cross it at night, and the mules coming behind would not be able to reach camp before morning. The argument was not

[1] A stringed instrument.

very convincing as we had crossed over just such a bridge only a few minutes before arriving at the camping ground, but I saw that further discussion was pointless since all the mules had been unloaded and turned loose to graze, and the loss of a mile was a minor debit entry in the total account for the day.

I found that only one complete tent had arrived. The poles and pegs of the others were among the luggage left behind at Manali. Luckily the *chholdaris* were available (the servants had been more provident than we) and I began to have these put up. The servants were tired and hungry, the luggage lay scattered over the stony ground in an untidy mess, just as the muleteers had thrown it off the backs of the mules. It was difficult to find anything. The sky was clouding over in a very threatening manner. A cold wind started and added to the confusion. As I was struggling with poles and ropes, Pal arrived, and, a little later, the Bhavnanis. They were not at all pleased with the first day of the trek. It had been hard going, riding on their small uncomfortable ponies, and the awkward gait of their mounts had given them aches and pains in strange places. As they sat on crates resting their tired limbs, they looked unhappy and forlorn. I tried to cheer them up by saying that the tents would soon be up. Mrs. Bhavnani, in a very tired voice, advised me to conserve my energies, and not go about exerting myself unnecessarily : 'Let the servants put up the tents; they can do it.'

A few drops of rain fell. It was getting darker and colder every moment. I had one of the Petromax lamps lit. A party of mules arrived, but these were ot the forty which had left Manali in the early afternoon

before we did. There was no news of Shrinagesh and
Bachittar Singh. I pulled up the zip of my jacket, and
sat down on a box.

'Dear God,' I said, 'don't let it rain to-night.'

At a quarter to nine I sent two men with a Petromax
lamp back along the path to meet Shrinagesh, and light
the way for him. Somebody had lit a fire and I boiled
a glass of milk and drank it. Then I made my bed in
one of the *chholdaris* and crept into it. As I lay in the
half-state between waking and sleeping I wondered if
the *chholdaris* would stay up till the morning. I had done
most of the work in putting them up myself, and I
hoped the pegs were firm in the ground, and the right
pegs had the right ropes tied to them. But my bed was
warm and comfortable, and induced a sense of indifference towards all that was outside it.

At eleven o'clock I was woken up by the dazzle of a
hissing Petromax lamp, and saw Shrinagesh holding up
the flap of my tent. I got up to speak to him. He was
tired and cross. He said :

'This is going to make us lose a day.'

He enquired about his tent. I asked him to come in
with me. He went away without saying a word. A
moment later I heard the sound of several voices and
the confused noise of things being moved and dragged
about, followed by the hammering of tent-pegs. Shrinagesh was having his tent put up. I turned over and
went to sleep. There would be time enough to hear
Shrinagesh's story in the morning.

The morning brought another day. I felt rested and
pleased with things. The tribulations of the previous

day had receded into a mist of semi-oblivion which invested them with the romance of mountains and sylvan glades, of rushing torrents and caves of ice, of the simple fare of shepherds and their charming flocks. These, in retrospect, seemed more beautiful and induced a greater measure of sensuous delight than when my limbs were tired and my senses were dulled by a long and wearisome march.

The tempo of proceedings in the morning was slow. No one was up till half-past seven. The first thing we did was to have the mess-tent put up and the radio unpacked. While breakfast was cooking we listened in to the 8.15 news, broadcast from Delhi. It came through, clear and unadorned with atmospheric embellishments. At breakfast, which consisted of porridge, fried sausages, the remnants of yesterday's tea, coffee and a fresh mango, we heard Shrinagesh's story.

A party of ten mules had arrived at four o'clock. These were quickly loaded, and the few things which were left over were entrusted to half a dozen coolies. The camp-beds had proved a source of trouble. They had solid non-folding frames and were too long for mules to carry over narrow paths ; so extra coolies had to be hired for them. Shrinagesh and Bachittar Singh were finally able to leave at five o'clock. They walked at a brisk pace, but before they had come to the end of the steep climb it became dark. Thereafter they had to proceed with caution, picking their way through the dark forest with care and circumspection. Often they stepped into channels of water which I had hardly noticed during the day ; once they lost their path and nearly went down a steep precipice; they waded through pools of mud and numberless streams ; the hillman who was acting as their guide was very little help, for the dense forest and a cloudy sky made every landmark invisible. Crossing the narrow pine-wood bridge was a nightmare. They had to get down on their hands and knees, and feel their way across the horrible chasm underneath, inch by inch while their senses were completely possessed by the roar and gurgle of the torrent below them. A hundred yards from the camp, they came upon the men I had sent with the lamp. They had gone just far enough to be out of sight of the camp and sat down to wait for the Commissioner Sahib !

I asked about the mules and the coolies and about Dr. Massey. 'They decided to camp when it became dark. They are only two or three miles from here, and should be arriving any moment.'

The tinkling of bells was heard and, almost at once, we saw a train of mules approaching. Dr. Massey joined us, and while his breakfast was preparing he began a recital of his misfortunes and tribulations. I regret to record that we were not very sympathetic. After consuming a substantial quantity of food, he went off to a lonely corner and, seating himself upon a wooden box, began to read from a pocket Bible and digest his breakfast.

We decided not to cross the Hamta pass that day, and make the journey to Chatru in two easy stages. This would enable us to organize everything in a more satisfactory way, and the loss of a day at this stage would save us trouble and delay later on. We sent back some of the *chholdaris* and camp-furniture. The beds were retained for one more day in deference to the wishes of Mrs. Bhavnani, who protested she would not and could not sleep on the ground. 'If one bug bites me,' she said, 'I lie awake all night.' But very soon we all learnt to sleep on the ground and never once missed the unwieldy camp-beds.

We started a few minutes before one o'clock, after the last mule had left. Our route took us through a beautiful valley, rising gradually to the foot of the Hamta pass. The path was rugged and, in places, steep. The Hamta torrent was now on our right, and on our left rose rocky precipices almost perpendicularly to a height of several thousand feet. We passed a number of snow bridges over the Hamta where the torrent rushed into a dark cavity formed by an archway of glistening ice, and rushed out again into the daylight fifty or sixty yards lower down.

The trees began to thin out, and as we approached the tree-line, which lies at an altitude of 11,000 feet, we saw clumps of rhododendrons bearing red, white and mauve blossoms, and *bhoj* trees (*Betula utilis*). Across the valley grew *devidars* (*Cupressis torulosa*), slimmer and smaller than deodars, but not less picturesque. We rose above the trees, and came to the region of grass and flowers. There were wild strawberries everywhere, and violets, anemones, forget-me-nots, buttercups, king-cups and a lovely red flower with velvety petals that I could not identify. The dandelions were larger and of a brighter colour than those that grow in the plains. Narcissi grew in abundance, but they were not yet in flower. I was greatly surprised to see the cobra-lily growing in several places. It reared its head in a most menacing attitude, and with its bright speckled stem and hood and its long sharp tongue, it looked far more realistic than its somewhat anaemic and insipid brother of Simla. Are there any cobras in these parts — at a height of 11,000 feet above sea-level ? If not, why this pointless imitation ? I asked the mule-men if they had ever seen any snakes in the Hamta valley. Their answer was an emphatic no.

We saw a family of Spitials camping by the wayside and stopped to speak to them. They did not understand Hindustani, but the man showed us his prayer-box, made of ornamental brass, containing an almost faded photograph of the Head Lama of his monastery, and a holy book, beautifully inscribed on leaves made from the bark of the *bhoj* tree. A little farther we met another party of Spitials who said they had come over the Hamta pass, and that the way was quite safe for pack-animals. We also met the mail-runner, an employee of the Nono, bringing mails from Spiti.

We traversed a long snow-covered slope and came to *Balu ka Ghera* (the den of bears), a flat grassy plain on the bank of the Hamta stream. Our men were already busy putting up tents. The kitchen fire was burning, and in a few minutes we were sitting down to tea. Joseph was putting up the poles of his wireless transmitter, and everything had an orderly and agreeable appearance.

Our camp lay at the foot of the climb to the Hamta pass. We could not see the pass itself, as low clouds hung over and around the peaks on all sides, but the immediate prospect was delightful. The infant Hamta stream, not yet grown to the vigour and turbulence of youth, meandered playfully through the valley in a series of wide curves. I could see the large U-shaped expanse of snow where it took birth, and the steep rocky precipice rising on one side. Altogether it was a delightful spot, and I wished the weather had been more favourable for taking photographs.

After tea we sat and chatted. I gave a demonstration of yoga exercises, and explained how a yogi could quench his thirst by merely breathing in a certain manner. It was a piece of pure exhibitionism, but it was justified by the interest displayed by everyone. Dr. Massey, in particular, was very attentive and asked me if it was correct that the practice of yoga enabled one to live without food for long periods.

I said : 'The practice of advanced yoga certainly does.'

I did not realize at the time that my light-hearted reply would cause a minor crisis on the following day, and lead to even more serious consequences in the days that followed, culminating in the eventual departure of Dr. Massey from government service.

But that evening we felt pleased with ourselves. The day's march had been short and extremely pleasant. We were camping in delightful surroundings. We had organized ourselves better, and the camp routine was smoother. Ashok had recovered completely. He had removed his bandages and had been able to walk part of the way — an achievement of which his mother felt justly proud. As we went to bed we saw that the sky had cleared, and stars were twinkling. We hoped that Indra, the god of rain, would deal with us kindly when we crossed the Hamta pass in the morning.

I was up at half-past three and we were able to make an early start. The sky was overcast and threatening. Indra was not willing to be clement. A fine drizzle began soon after we started, but the rain held back its violence till we reached the top of the pass. The ascent was not difficult, but was certainly what I should call interesting. The route lay entirely over snow-covered slopes of gradually increasing gradients — the last few hundred yards being very steep and slippery. The ponies found the going hard and tricky, and the riders had to dismount. Before starting in the morning I had taken the precaution of unpacking my *poolas* (rope-soled hill shoes) and putting them in my haversack with the cameras. I took them out now and wore them in place of my shoes. I found that they gripped the semi-hard snow much better than my leather soles, and enabled me to walk faster and with greater confidence. My hobnailed *chaplis* would have served me even better, but a broken strap had rendered them unserviceable for the time being.

We reached the top at half-past nine. It was raining hard and the air was cold and biting. There was much less snow on the other side of the pass, the path being completely free of it, but the heavy downpour, combined with the steepness and narrowness of the track, made the downward journey a difficult one for the mules. They slipped and slithered on the wet rocks, and had to be coaxed and encouraged at every step. One of them stumbled and fell down. It had to be unloaded, taken down a short distance and then reloaded. The tents and beddings were getting wet and heavy. The mule-men sulked.

In these conditions our cook, Chanchlu, snatched an opportunity of fulfilling himself and involving me in a few moments' uneasiness and a painful experience. One of the mule-men had brought with him his wife, a young woman of not unpleasing countenance who was as strong and full of vigour as any of the men. She loaded her mules like any man ; carried heavy packages and raised them up to place them on the back of the mule with surprising ease. While on the march she always led the way and seemed quite tireless. She was among the first to reach the top of the pass, and I was surprised to see her standing below an overhanging rock away from the path and almost invisible from it. The mules had already gone down and she could not be taking shelter from the rain at such an unlikely place ; besides, the rain gave no indication of ceasing for the day, and it would have been foolish to seek temporary protection from it. But when I saw Chanchlu crawling up behind me with his ant's legs splashing through mud, I understood the significance of her strange conduct. Anxious to avoid any suspicion of being considered an eavesdropper, curious or censorious, I hurried on and took a narrow footpath which I thought would serve as a short-cut and enable me to overtake our main party. The footpath brought me to the edge of a steep slope completely covered with hard snow, and ended there. The grass near the edge of the declivity of snow was wet and slippery. Even my *poolas* could not keep a hold on it. I tried the snow and narrowly escaped a danger-ous fall to the bottom. By now the footpath was several yards above me and seemed out of reach. I said a great many things about Chanchlu and the muleteer's wife, and sat down to think. Suddenly I had a brain-wave. I pulled my waterproof cape tight around me and gathered up the lower end by clasping my hands round my thighs below the knees. Squatting in this position, I raised my heels from the snow, and began to slide down the hill. If I dug my heels in I could stop in the space of a few feet. Thus, slowly at first, and faster as I gained confidence, I travelled down the hill for a distance of more than three hundred feet, and reached the bottom in two or three minutes. Before me lay the path and I saw Shrinagesh coming round a bend. He had been worried and was about to send a man to look for me when he saw me glissading down the snow slope.

We walked together the rest of the way, plodding through rain and slush. Neither my 'surplus army stores' waterproof cape nor Shrinagesh's Burberry afforded any protection from the torrential downpour to which we were being subjected. Streams of water ran down our collars, soaking our very underclothes and chilling every limb of our bodies. Our trousers clung and flapped about our legs in heavy, clammy folds. The rain flew off our sleeves and hands in a continuous shower. It was a strange experience, this feeling of intense wetness. There was a sense of having been merged with the elements, of having lost one's individuality, and being a part of the rain, the wind, the cold and the mud, and not something apart and distinct from them. It was as if our bodies had dissolved into the rain and the wind and only our disembodied spirits were feeling the impact of pain and discomfort. After a time even this feeling disappeared, and as the senses became numbed, all that remained was a consciousness that there were cold, piercing spots where one's knees, shoulders and hands used to be. We walked because there was nothing else to do. We hardly spoke a word, because there was no topic that required discussion. We had no grievance and we made no complaints. We went on and on, down the narrow path, occasionally uttering a warning sound, 'Careful', 'Slippery here', 'Mind your step', till we reached the camping-ground at Chatru, and sat down on a large slab of stone.

It was half-past twelve on July the 4th, and the rain was pouring as I have seen it pour in Simla or in Delhi during the worst period of the monsoon. Had we really crossed the Hamta, and did the Great Himalayan Range lie behind us? Or had we not yet come to the region which is beyond the reach of the monsoon? The barren rocks staring at us from all sides, the tall bleak mountains standing over us and crushing us from both sides of the valley, the angry torrent raging over stones and boulders a hundred feet below us, the fierce cold wind which lashed the rain about us and went howling through the valley, told us that we were in the heart of the Himalayas and face to face with Nature in her cruellest and most relentless mood.

IV

THE CHANDRA VALLEY

THE Chatru camping-ground is a small grassy incline scarcely larger than a tennis court, sloping down towards the River Chandra. It ends abruptly at the edge of a steep declivity running into the torrent below. On one side of it stand tall boulders, more than fifteen feet high, and from the remaining two sides rocky mountains rise and soar up into the sky. Despite its sheltered appearance Chatru receives the full fury of wind and rain and our tents were in constant peril of being uprooted and blown away. We had to adopt the device of throwing *chholdaris* over our tents to weigh them down, but the tents shook and trembled all through the afternoon and for half the night, till the wind suddenly dropped, and a calm silence fell upon the air. This phenomenon is a peculiar feature of the inner Himalayas, and we had repeated evidence of it throughout the trek. The wind rises at noon each day, and blows vigorously till about midnight ; it then retires for the space of twelve hours.

It was not easy to put up tents in such a high wind, and the rain added to our difficulties. We had to struggle with ropes and poles, and wrestle with canvases flapping violently to escape from our hands, but we finally succeeded in putting up the kitchen *chholdari* and three of the smaller tents.

The rain stopped suddenly and bright sunshine filled the valley. We put our wet clothes out to dry and sat down to wait for lunch.

Dr. Massey's voice, loud and querulous, was heard above the noise of the wind. He was involved in a vehement argument with someone. Stray, unconnected words pushed their way across to our ears. The discussion, it appeared, was concerned with food, rain, being hungry, getting wet and a tarpaulin, particularly a tarpaulin, as this word was repeated several times. The voices approached and the two disputants — Dr. Massey and Bachittar Singh — stood before Shrinagesh, asking for an adjudication of their rights.

Dr. Massey said he had been soaked to the skin, had eaten nothing since six in the morning, had had to undergo unprecedented hardships, the journey up to the Hamta pass had been beyond his powers of endurance, his heart and lungs had all but refused to perform their functions, the cold wind had frozen the blood in his veins and paralysed his senses. He had not been able to find the ground-sheet of his tent, and had, therefore, appropriated a large tarpaulin which no one was using but which was claimed by Bachittar Singh as his property and his protection from damp and cold.

We were informed, on the other hand, that Dr. Massey had ridden a pony all the way from Balu ka Ghera to Chatru, that he was no wetter than the others, that he had sneaked into the kitchen tent and prevailed upon Chanchlu to serve him with a substantial meal, and that the incompetence or reluctance of his heart and lungs could not be pleaded in justification of taking someone else's tarpaulin.

Shrinagesh appealed to me : 'This looks like something you ought to handle.'

I tried to reason with Dr. Massey. I agreed that it had been a long and trying day for him, but had we not all suffered in equal measure ?

'Look, Dr. Massey,' I said, 'you had a pony to ride all the way. I walked, and I am older than you are. I had my last meal at six o'clock, nine hours ago, and I have had no tea or coffee since then. I couldn't, because the man carrying the thermos flasks stayed behind, with you. Perhaps you asked him to. Please don't try to make things more difficult than they are. Now, be a good chap and give Mr. Bachittar Singh his tarpaulin. It belongs to his tent.'

Dr. Massey's cup of resentment was full. It flowed over in a loud and violent invective against all of us. We had not treated him like a human being. He belonged to a subordinate service We thought ourselves the 'heaven-borns'. He might have foreseen what would happen to him if he travelled with I.C.S. officers. He was completely at our mercy. We could throw him over a precipice, leave him behind to die, starve him, kill him, and nobody would so much as question us. Was it his fault if he was hungry after a whole day's long and tiring march ?

'But we are all hungry,' I said. 'We haven't had lunch yet.'

'You do yoga exercises. You can do without food. You said so, yourself, only yesterday.'

He said this in a tone of such tragic resignation that we could not help laughing. Fortunately lunch was served at this moment, and Dr. Massey sat down to make good any deficiency he might have suffered during his hurried and surreptitious visit to the kitchen-tent.

The tarpaulin was restored to its rightful owner and Dr. Massey sat down near a large boulder to sun himself and read the Bible. But he was never the same man again. He kept even more to himself than before, and seemed to spend a great deal of time in deep meditation. Subsequent events suggested that he had, from that moment, been ready to take the earliest opportunity to avenge himself on us, and show his disregard for authority. For the moment, however, peace reigned in the camp and the sun was warm.

Shrinagesh summoned the muleteers and questioned them about our route to the Kunzum pass. Were there any large streams or dangerous torrents to be crossed ? Were there bridges over any of them ? What did they know of Shigri ? Was it fordable ? Was it safe ?

Everyone had a different story to tell, and those who had never before come this way were the loudest and the most positive in their assertions. We, however, noted a certain measure of unanimity on some points : the path going up to Kunzum *la* [1] was good but very steep ; we should have to cross several streams ; there were no bridges over any of them, but most of them were easily fordable ; it was advisable to make an early start, and cross the streams before eleven o'clock, as some of them assumed unfordable dimensions as the advancing day melted the snows that fed them ; the Shigri was the most troublesome of them all, because of its capricious

[1] *La*, Tibetan for 'pass'.

nature ; sometimes it spread out into a wide and shallow delta when a child could walk across it, at other times it gathered up its waters into one fearsome torrent when not even elephants dare venture to step into it. When this happened one had to leave the valley, climb up the mountainside to a point beyond its source, and traverse the glacier — a tricky business. A former S.D.O., a rash young Britisher, one of Bachittar Singh's predecessors, had nearly lost his life while crossing the Shigri.

Our journey to Spiti promised to be interesting.

The wireless operator was able to get through to Jullundur, and he brought a sheaf of messages for Shrinagesh from the Superintendent of his office. Nothing startling had happened, and the civilized world was pretty much where we had left it three days ago.

As we went to bed, I saw that the sky was bright with stars and the Milky Way stretched across the heavens, luminous and glowing like a huge fluorescent band. We had decided upon an early start, and as I wrapped the bed-clothes around me I whispered to myself : 'Wake up at four o'clock.'

I was woken up by the sound of rain-drops pattering on the roof of my tent, and by a feeling of strange discomfort in the lower part of my body. It was some time before I realized that the source of this discomfort was no other than the pull of the earth. It was gravity made manifest by the unusual situation of my bed.

What a terrible and disconcerting power is the force of gravity. The simplicity of the mathematical formula, $g = 32$ ft. per sec. per sec., gives no clue to the ubiquitous and pervasive nature of this influence, nor does it contain any indication of the complexity which it introduces into all spheres of human conduct. If it protects and preserves us, it also annoys, tantalizes and frequently defeats us. It keeps our bodies and even our minds tied down to the earth. I remember as a child going with my mother to visit a sick friend of hers. As we entered a room full of chattering women we saw the invalid lying on a bed of which the foot end had been raised by placing a brick under each leg. I asked my mother the reason for this extraordinary state of affairs. She 'shushed' me into silence, but I heard the loudly whispered talk of the women-folk, and connected an occasional word with things I had heard the boys speak of, in remote corners of the school playground. On leaving the house I again questioned my mother. She told me the lady was sick and tired, and unless her feet were raised higher than her head, the blood rushed to her feet and made her more sick and more tired. I fastened on the word 'blood', and wanted my mother to give verbal shape to the incomprehensible idea teasing my mind, but she would not assist me further.

In later years I learnt to recognize and accept, with resignation, the devastating nature of the attraction which the earth exercises upon us. I heard of the apple which fell on Newton's head, and of Galileo's pendulum. I saw that the power which keeps the universe together and the planets revolving in their orbits also makes us old. It gives us drooping lips and sagging bellies. It is responsible for double chins and pouches under the eyes. The triumph of gravity implies bondage, unhappiness, sickness, depression. All the vocabulary of pain and sorrow is derived from this source : Hell is the *nether* region. I am feeling *low, down* in the dumps, I have gone *under*, I have reached *rock-bottom*, the *lowest* point, the *nadir* of unhappiness. I am in the *depths* of depression, my heart is *sinking*, etc., etc. . . .

It was gravity which brought the rain beating down upon us the day we crossed the Hamta, and again during the night, wetting our tents, and it was gravity which made me conscious of the congestion in my legs as they lay sloping down the inclined plane. I began to battle against the tendency to slip down. It was one o'clock by the luminous dial of my wrist-watch. The day and its labour were still three hours away and I must have my full measure of sleep. If only I could draw the blood back from my feet and legs, if only I could stop myself from sliding into the River Chandra, if only I weren't falling into the abyss below me with an acceleration of 32 ft. per sec. per sec., if only . . .

My first thought on waking up was that the tents were wet and heavy, and the muleteers would make a fuss about carrying them. We should not be able to start till late, and the streams would be swollen and unfordable.

The rain had stopped, but a mantle of dark clouds lay over the valley. Some of the tents had been taken down and packed. The mule-men patted the soggy bundles, turned them over, lifted them a few inches off the ground, jerked them up and down and dropped them. They shook their heads, and said the mules could not possibly carry loads of such crippling magnitude. The tents were unpacked and spread out to dry. At six o'clock the sky cleared and a bright sun began to shine. At seven the mule-men conceded that the tents were sufficiently dry for transport. We began packing at once and were able to leave at a quarter past nine.

The path went down to the bed of the Chandra river and proceeded along its left bank·for almost a mile. We came to a small stream which we forded easily by removing our shoes and socks and rolling up our trousers. The water was cold, but not deep, the stones underfoot were slippery and insecure, the smaller pebbles had not yet had their corners and edges rounded off and smoothed, but I found the experience pleasing if only because of its novelty. At another stream I watched a flock of goats crossing over. Their instinct led them unerringly to the safest place, and the goatherd followed. It was clearly not a question of memory, and a previous year's crossing would be of no assistance in locating a safe point of transit, because the continuous movement of stones and boulders, brought down by the force of the torrent, alters the character of the stream from day to day, and sometimes it is impossible to know just where the water is shallowest, and where the bed of the stream is free from treacherous holes. The local people have implicit faith in the judgment of animals, and allow them to choose their own path across a stream. We saw our mule-men doing the same, and more than once I was amazed to see the leading mule pause near the edge of a stream, sniff at the water and, after going up or down the bank a few yards, enter the water deliberately, and walk across to the other side. The remaining animals followed close behind. The muleteers always advised us to ford the stream at the same spot. If a flock of goats or sheep came to one of the deeper streams, they climbed up the mountainside till they came to a point where the crossing could be safely effected. The *gaddi* followed, whistling encouragement or approval, as the occasion demanded.

We came to a stream in which the water was about three feet deep and the current fast. The mules and horses were able to ford it, but we were advised to go a little way up the hill and cross the stream by a snow bridge. We walked over a magnificent field of snow, more than a hundred yards long. A deep crevice lay gaping, half-way across it, and at the bottom of the chasm a torrent roared and swirled like a mass of liquid boiling in a cauldron. The final span of the bridge was a huge arch of solid snow ten feet thick, with a gap in the centre two feet wide, of which the ends faced each other like two halves of a gigantic vice. We jumped across the gap and stood for a long time looking at the breath-taking beauty of this scene. In a few days the gap would widen further and the snow bridge would no longer be a bridge.

The path now left the river-bed and climbed up to a beautiful grassy plateau through which little rustling rills meandered. Some goatherds were camping here and we met the local *zaildar* [1] who had come from Spiti to welcome the Commissioner Sahib. The *zaildar* undertook to construct a bridge over the *Chhota* [2] Shigri stream, and went ahead to make the necessary arrangements. The undertaking was carried out to our entire satisfaction ; and the next morning when we came to the *Chhota* Shigri we found an excellent bridge made

[1] Village official.
[2] *Chhota*, small. The *Bara* (big), the real Shigri, lay farther on.

from a pair of pine spars upon which a number of flat
stones were resting. These were reinforced and surfaced with handfuls of grass and moss. The mules
approached and sniffed at this contraption very suspi-
ciously, but when the first one had been coaxed and led
across it in safety the others followed quietly, inhaling,
at each step, the reassuring odour left by the previous
traveller all along the perilous route. It was a good
piece of work and the *zaildar* showed great organizing
ability in procuring pine spars in a treeless tract and
labour to build the bridge in a completely uninhabited
area. I never found out how he did it.

The Chandra valley has a rugged grandeur which
at once inspires awe and admiration. It frightens you
by its relentless power, and yet you are fascinated by its
wild beauty and drawn towards it. The Chandra rises
from a huge snow-bed lying on top of the Baralacha
pass. A mile from its source it is already a considerable
torrent and quite unfordable. For sixty miles it rushes
madly through a narrow barren valley in a wide curve,
taking continuous tribute from numerous glaciers all
along its course. The slopes on either side are too steep
and rugged to permit the growth of any vegetation, and
this part of the valley has no villages or habitation of
any kind. Here and there lie grassy pastures, perched
above great heaps of fallen stones and debris, and to these
pastures *gaddis* from Kangra and Kulu bring their flocks
of goats and sheep for a few months' peripatetic holiday
each year. They come in early June, make a round of
the various pasture grounds, and return home before the
September snows close up the passes. The picturesque

camping-ground of Shank Shum where we met the *zaildar* is one such place, the broad grassy bowl round the Chandra Tal is another. We saw several more, but these did not reach the same standard of beauty or richness. Except for these rare oases, the valley is nothing but rock and snow moulded in a hundred different shapes and colours until Khoksar, the first village, is reached. From here the valley widens out and is dotted with little hamlets and villages whose inhabitants eke out a scanty living by growing barley and buckwheat.

The Chandra remains throughout at an altitude of more than 9000 feet. At Shank Shum we were nearly 12,000 feet above sea-level.

As I continued my way along the stony path beyond Shank Shum I found the going hard and sat down to rest on a flat rock. Thick dark clouds were massed above the peaks on either side of the valley, and while I watched them they began to move across menacingly. The moment they entered the valley, they changed to a lighter colour. Little wisps and shreds broke away from the main body, and floated into the sky directly overhead. The stray pieces became fleecy and in a few seconds completely dissolved into the air and disappeared. The sky above the valley, along its entire length, remained clear and of a deep, deep blue. It was a remarkable phenomenon and I watched it, lying on my back, fascinated for nearly half an hour.

I tried to find an explanation of this extraordinary behaviour of the clouds which I had not observed elsewhere, and it seems to me that the explanation is this. The midday sun plays upon the snow, piled up on the high peaks on both sides of the valley, and in the intense heat (for there is very little absorption of the sun's rays in the rarefied atmosphere of 20,000 feet above sea-level), the snow melts and evaporates. The vapour, chilled by air, condenses into thick clouds. At the same time the air at the bottom of the valley is warmed and, becoming lighter, rises. A vacuum is thus produced directly above the valley, and clouds from both sides move across to fill this vacuum. As they leave the cold snowy peaks they are met by an upward current of hot air, and begin to disintegrate with its impact. The clouds break up, and the condensed vapour once again evaporates. Above the centre of the valley the air is hottest and no trace of condensation remains.

A similar observation of cloud behaviour in Spiti confirmed the correctness of my hypothesis. The phenomenon I observed is met with only where tall snow-covered peaks enclose a narrow barren valley — the snow provides the moisture for the clouds and the narrow valley generates a column of hot air which continuously rises and dissolves them. But whatever the scientific explanation, the spectacle of thick dark clouds lowering from the mountain-tops, and then rapidly melting away into nothingness is something incredibly fantastic, and must be seen to be believed.

When I started again I found myself alone. As far as the eye could see there was no other living being in the whole valley. Our mules and servants had gone on ahead, and a few stragglers had passed me as I lay watching the clouds. The man carrying our lunch had stopped and given me a cold *paratha* [1] and some baked beans congealed into a solid clod. The *paratha* was like

---

[1] Unleavened fried bread.

a piece of dried leather, and the beans were an insult to the very name of Mr. Heinz. The combination was singularly distasteful even to the palate of a tired and hungry traveller, and I was left with a feeling of general dissatisfaction. I realized that, in all probability, I was the last of our party, and hurried forward. The path I was following could scarcely be called a path at all. It was a vague ill-defined track that disappeared for long distances in the confusion of stones and debris that lay scattered all over the valley. At one place I lost the track completely and climbed up the hillside on the right. It was a foolish thing to do, for very soon I found myself in a veritable forest of stones and boulders, which lay packed closely together all around me. They made progress in any direction extremely difficult and shut out the view from all sides. I had to jump from stone to stone, and keep my eyes searching for the next precarious foothold. In trying to extricate myself from this horrible wilderness I only became further involved in the intricate pattern of rocks and boulders. The exercise was tiring my limbs and my spirits. I fell down between two stones and bruised my knee. I noticed a small tear in my trousers. It was dreadful to be alone in this vast cruel valley, out of sight, out of hearing of anyone. I had heard someone say that the local people do not venture into this valley singly. They always go in twos or threes, even the regular mail-runner takes a companion with him. I realized that if I sprained an ankle and became incapable of hopping over these boulders I should have to stay there for ever.

I climbed down the hill slowly and regained the river-bed. A hoof-mark restored my equanimity, and a few yards farther fresh mule droppings indicated that I was on the right path. Two hours later, tired and hungry, I arrived at the camping-ground of Phuti Runi, where our tents had already been put up.

Phuti Runi means cracked or broken stone. The stone in question is about thirty feet high. Its dimensions at the base must be forty feet by thirty feet. It lies in two parts with an open fissure all the way down. The fissure was obviously caused on a cold and frosty night when rain water which, during the day, had entered a narrow crack froze and expanded, blasting the two halves of the stone apart. There must have been a tremendous explosion when the stone broke, and the noise must have gone reverberating through the Chandra valley for miles. The cracked stone is an impressive sight and a monument to the terrible forces of nature. By its sheer size alone it made one of our muleteers standing near it look like a puny insect.

The Phuti Runi camping-ground, with its atmosphere of expansive peace, is almost as delightful as the pasture ground of Shank Shum. It is broad and roomy and is carpeted with a thick pile of soft green turf. On one side of it lies a low ridge beyond which a lovely stream of limpid fresh water flows through a meadow strewn with dandelions and daisies. In this stream we bathed. The water was ice-cold, but we lay basking in the hot afternoon sun and felt greatly refreshed after the day's tiring march.

The next morning we were up at half-past four and on our way at a quarter past seven. After crossing the newly constructed bridge over the *Chhota* Shigri we concentrated all our thoughts on *Bara* Shigri — our *bête noire.* We had sent a P.W.D. mate and two men with pickaxes to prepare a way over the ice, for we had abandoned the idea of even making an attempt to ford the capricious torrent. The path (if it can be called that) left the river-bed, and went steeply up the hillside on our right. It was very difficult going ; we had to step from one boulder to another and, every now and again, we stopped to look around and see where we had to go. It was almost a repetition of my experience of the previous day except that now I was not alone. Occasionally we saw a couple of stones, piled on top of a boulder. This was a mark left by the shepherds in their passage to indicate the route for the benefit of those who would follow ; but since similar marks placed in former years had not been removed, and the glacier keeps continuously moving and altering its shape, the little piles of stones were sometimes misleading.

We toiled on through this labyrinth of rocks for more than a mile. On the way we met the P.W.D. mate coming back to tell us that he had been across the glacier and had found the path quite safe.

The Shigri glacier is a mountainous mass of solid ice, more than a mile wide, and extending over a distance of six miles from its tail at the top of the mountain at a height of 20,000 feet above sea-level, to its mouth at the bottom of the valley. At this point the Shigri torrent debouches from a dark cavern below a huge archway of iridescent ice, and rushes in swirling eddies to join the Chandra river. Large blocks of ice are detached from the walls of the tunnel under the glacier by the force of the water and carried downstream like miniature icebergs. The snout of the glacier is strewn with thousands of stones, blasted by frost action and hurled down from the mountain tops. Where there are no stones, there is a thick coating of dust and gravel over the ice. At one place I chipped off a piece of ice with the point of my alpenstock. It was clear and transparent like an ice cube from a refrigerator, and I sucked it to quench my thirst.

The passage across the glacier was strenuous and, in parts, difficult, but in no way dangerous. The mules had a very uncomfortable time, and had to pick their way slowly and cautiously through the stones. A false step must certainly have meant a broken or a badly damaged leg. The official gazetteer says that laden animals cannot cross the glacier. We had chosen to disregard this warning, and fortune had favoured our daring.

Shrinagesh and I were the first to cross the glacier, and we went ahead in our anxiety to cover as much distance as possible before lunch. The route, for some miles, lay over rough and stony hillocks and then descended to the bed of the Chandra. Another two miles and we came to a flat grassy patch on the river-bank. Here we sat down to rest and wait for lunch. We had to wait a long time. The mule-men arrived and said this was a better place to halt for the night than the camping-ground across the Karcha stream, where the grazing was not so good. We were quite agreeable to this suggestion, and not at all loath to be spared the labour of walking another mile and having to deal with the Karcha stream in the afternoon.

The tents were put up, the horses were turned loose to graze, and the mule-men began piling up their *numdahs* and saddle-bags to form a barricade against wind and rain, but there was no sign of the man with the lunch. We were told he was with the Bhavnanis, who had stopped to take photographs of the Shigri glacier. Shrinagesh foraged in the provision box and found a tin of condensed milk, a packet of biscuits and some fresh plums from Major Banon's orchard. We made a meal of these.

An hour later the Bhavnanis arrived, completely tired out. The day's march had been too much for all three of them, and their spirits were low. Mrs. Bhavnani had fallen off her pony twice while crossing the glacier and this had unnerved her. Bhavnani had been greatly impressed by the Shigri and had spent a long time taking movie and still shots of the glacier and of the torrent from various angles, but the journey had exhausted him. At one stage he had reached the condition in which he experienced acute discomfort on the back of his pony, but, if he dismounted and walked, the torment of exercising his tired legs was even greater. Mrs. Bhavnani remembered the treks she had done in Kashmir with a feeling of bitter nostalgia. She had travelled in such comfort then. The contractor had provided them with tents furnished with beds, tables, chairs, commodes and canvas baths in which you could lie down and wash away the day's fatigue. He had supplied four hot meals a day. The riding-ponies were real riding-ponies and not clumsy uncomfortable monsters who shook from side to side, at every step, without any sense of balance or rhythm till every part of the rider's body ached. Oh, they had travelled and lived in absolute luxury in comparison with what they were

F

undergoing now. It had been so cheap too ; once for Rs.600 per person they had done a trek of thirty-five days and covered over five hundred miles !

The man with our lunch had arrived, and at five o'clock we had a substantial meal. One of our men had shot a couple of snow-pigeons the previous day ; Chanchlu had roasted them, and their meat tasted delicious. We dispensed with dinner and retired at eight o'clock to be able to make an early start the next day.

We set off at a quarter past five in the morning and, in a few minutes, came to the Karcha stream. Bhim Sen, one of the orderlies, told us that the stream was not fordable, and we should have to go a mile higher up and cross over by a snow bridge. He assured us that he had been along this route before, and knew what he was talking about. So Shrinagesh and I, accompanied by the man who was carrying my cameras and our lunch (we had taken the precaution of having our food packed separately), followed the orderly. At the place where there should have been a snow bridge we found two walls of ice, with a breach twenty feet wide, through which the Karcha stream rushed out with a roaring noise. It was a beautiful sight, but we could not cross the torrent at this place. We turned round and, in the far distance, saw the caravan of mules crossing the stream at the point which Bhim Sen had pronounced impracticable. So we hurried back to where we had started from. We were not surprised to see that the volume of water had increased appreciably during the time we had taken to make the return journey to Bhim Sen's snow bridge. We removed our shoes, socks and trousers and entered the stream clasping each other's hands, to acquire greater stability and weight. The water came up to our thighs, the current was fast, and, as it pushed us about, our legs were chilled and the stones underfoot hurt, but we crossed over without any mishap. The Bhavnanis had gone across an hour earlier on their ponies when the depth of the stream was several inches less.

Leaving the Karcha we followed a narrow track scarcely six inches wide across the face of a steep moraine with its feet in the Chandra river. The path was so narrow and lay across such a steep slope that I dared not look down or back. A false step, an inadvertent slip must have proved fatal. There were no bushes, stones or irregularities to relieve the smooth face of the moraine and there was nothing to obstruct or break the downward rush of anything rolling down the declivity. If anyone slipped he would be precipitated into the river and be irretrievably lost.

We went up and up, panting in the rarefied air, the toil-drops falling from our brows like rain. It seemed the climb would never end. We despaired of seeing the Kunzum *la*. A sudden turn to the right brought the top of the pass into view. We climbed the last half-mile of the saddle, stopping to rest every hundred yards, and reached the top at half-past eleven.

We were in Spiti.

V

SPITI (1)

OUR first view of Spiti [1] was exciting and full of promise. From where we stood at the top of the pass the mountains fell away in a vast series of undulating slopes down to a valley, concealed behind a ridge which ran across the line of sight. Beyond the ridge, in the far distance, stood a mighty wall of crags and peaks, of weird shapes and fantastic colours, varying from deep crimson to slate and ashy grey. To our right rose two magnificent snow-covered peaks which, my map said, were 20,000 feet high. On our left the pommel of the Kunzum saddle went up an easy gradient to a large patch of snow. At our back were the lofty peaks and glaciers of the Great Himalayan Range, with their feet in the Chandra river, now invisible to us. The picture all around was one of desolate grandeur and magnificent splendour, but there was nothing terrible or fearsome in this scene. It was like an immense work of art created to please and convey the meaning of universal truth. One wanted to stay and drink in the beauty of it all and not shut one's eyes and cower down before it. That had been our reaction to the impact of

[1] Spiti, locally pronounced Piti, is a Tibetan word, meaning 'middle province'. It was so called because of its position between British India, Kashmir, Tibet and Bashahr.

the Chandra valley which inspires awe and fear by its
sheer sense of power and ruthlessness. From the Kunzum pass Nature did not appear hostile or angry. Perhaps our lofty position gave us a sense of confidence.
There were no bleak mountains or monstrous rocks
towering over us. The mighty Himalayas lay spread
out before us and below, in tier after tier, and stretched
away to a far horizon. A bright sun shone out of a
deep blue sky and the air was almost still. Over everything lay a brightness and a quiet peace which at once
excited and soothed the senses.

There was a cairn of stones at the top of the pass with
hundreds of prayer-flags fixed all over it, giving it a
gay and festive appearance. There were stones with
beautiful carvings arranged on top of the cairn. The
carvings and the inscriptions on flags said, *Om manipadme hum* (Hail, thou possessor of the jewel lotus.
Amen) or some other text from Buddhist scriptures.
They had been placed there by way of thanksgiving for
benefits received, or as prayers asking for a blessing and
future favours. A traveller who has successfully braved
the rigours of a journey over perilous paths, and has
safely arrived at the top of a pass says, *Om manipadme
hum*. These words bring peace to his troubled soul and
give him courage in his future ventures.

This *mantra* or spell is a 'saviour of all beings from
all dangers, a releaser from all pains. It is like a pool
for the thirsty, a fire for the cold, like a garment for the
naked, a caravan leader for merchants, a mother for her
children, a ship for those sailing across, a physician for
the sick, a lamp for those wrapped in darkness, a jewel
for the seekers of wealth, a universal King for princelets,

an ocean for rivers, a torch for dispelling all darkness.'

The words of the spell are often misinterpreted. They have been variously translated as, 'Hail, the Lord is in the Lotus', 'O, precious Lotus, amen', 'O jewel in the Lotus'. But *manipadme* is one word, substantive, feminine gender, vocative case, derived from *manipadma*, which means, 'she who has a jewel lotus'. The person addressed is a female deity. Buddha is said to have learnt the spell after a great deal of toil and devotion. But when a *bodhisattva* or seeker of *Nirvana* asked him for it, Buddha sent him to a reciter of the doctrine at Benares. When the reciter was approached, a voice from the sky ordered him to communicate the *mantra*. It was the voice of Avalokitesvra, the lord who surveys the whole creation. The *bodhisattva*, having learnt the spell, returned to Buddha and seven hundred and seventy million Buddhas assembled and repeated the words *Om manipadme hum* — Hail, thou possessor of the jewel lotus. Amen.

For the modern Buddhist of the Mahayana school, it is not enough to say the words orally. He writes them on a piece of cloth or inscribes them upon a stone and leaves them at the spot where he pronounced them with his lips, so that they will go on echoing sempiternally, and, by the power inherent in their sacredness, continue to bless him and watch over him. These monuments, *mane* walls or *chortens*, are to be found everywhere in Spiti and Lahoul — near villages, on top of passes, on the bank of a torrent, near fords and the confluence of rivers. Some of the stones have a beautiful red, black or green colouring, and the carvings are executed with great skill and craftsmanship.

There must be millions of such stones in Spiti, for these *mane* walls are four to five feet high, three to four feet wide and several yards long. We saw one more than a hundred yards in length. The visitor is tempted to pick up one of these stones and carry it away as a memento of his visit. But woe betide him if he does, for ill-luck will pursue him in one form or another till he restore the stone to the *chorten* of which it formed part originally. So it is that through the centuries these *mane* walls have subsisted and gone on augmenting, and with every breath of air they whisper the sacred words of peace : *Om manipadme hum*.

We too pronounced the magic words and started on our downward journey to Losar. It was easy going after our strenuous climb, and the gradient was gentle. What wonderful slopes to ski on, I thought. If only one could go there in winter when all the mountains and valleys are covered with snow, and the wrinkles and creases are smoothed out. One could go on and on for miles, skimming over soft, rounded billows of powdery snow.

We came to a small enclosed valley through which a tiny stream went bubbling and rippling on its way, in slow winding curves, over a bed of red stones, like a rivulet of red wine. All around lay a carpet of thick green grass with a pattern of multicoloured flowers. There were bunches of forget-me-nots in blue, pink and bright yellow, white and purple anemones, butter-cups, daisies, wild peas, iris, Michaelmas daisies and edelweiss. The edelweiss was small and somewhat wilted, but as this was the first time I had seen this plant growing in India I picked a few flowers and pressed them in my notebook. Later I collected some beautiful specimens with large velvety petals of a silver-green colour.

We came to a large stream spread out in several small channels over a wide stony bed. The water was shallow and not more than a foot deep ; it was too much trouble taking off our shoes and socks and putting them on again each time we crossed one of the small channels, and we began to walk through the water. We were about half-way across when I saw a band of queer-looking horsemen hurrying towards us. They were men from Losar — the headman and three others who had been waiting for news of the Commissioner Sahib, and, with the arrival of the first batch of our mules, had set out to meet Shrinagesh and escort him to their village with due honour. Their features had the peculiar Mongolian cast — narrow eyes and high cheek-bones. They were ruddy-faced and short-statured and strangely attired in a long, loosely fitting double-breasted coat of coarse woollen material, girt at the waist with a long woollen rope, wrapped round and round in multiple coils. Their headdress consisted of a tall felt hat with turned up embroidered rim. They wore their hair well greased and done up in a single plait which hung below the neck, and by its continuous swinging motion plastered the back of the coat with a triangle of thick shiny grime. All around them they poured out an abundance of fulsome odour, and spoke a language we did not understand.

It was, however, obvious that they were welcoming us to their land and asking us to ride their horses, at least while fording the stream. They were greatly

distressed at the sight of our wet shoes and trousers
and shook their heads sorrowfully, as if some dereliction
of duty on their part had occasioned our sorry plight.
It would have savoured of peevishness if we had insisted on adhering to our resolve to walk all the way,
and we had not the heart to add to their discomfiture
by refusing the offer of their mounts. Besides, what
would they think of a Commissioner and a High Court
Judge who had to trudge like common men even while
crossing a stream? So we struggled up and into the
uncomfortable high-pommelled wooden saddles, and
allowed them to lead us across the water.

After crossing the stream we dismounted and walked
till we reached Kala Khol (the Black Stream), so called
because of the dark grey, almost black, colour of the
soil at this place which makes the water of the stream
look inky black from a distance, and also because of the
treacherous nature of the torrent which, every afternoon, brings large boulders hurtling down its steep
course with a horrible, ominous roar. We had been
warned about Kala Khol, and were glad that all our
pack-animals had gone across safely.

Once again we sat perched on the wooden saddles
and were being led across the stream. Shrinagesh was
ahead of me. The man holding the reins of his horse
jumped from one boulder to another, fell into the water
where it was shallow and scrambled on to another
boulder. He dragged his horse behind him, regardless
of the fact that the poor animal could not hop from one
boulder top to another. Shrinagesh was powerless ; he
said he would guide his mount himself, but his protests
went unheeded. Perhaps the man took them for words

of praise and encouragement. The horse stumbled on for a few paces, diving into deep water and scrambling over rocks. Shrinagesh rocked forward and back again, clutching the pommel of his saddle. In midstream the horse slipped and nearly fell on his side. He leaned against a rock, jumped up and bounded out of the water, taking Shrinagesh with him.

My horse followed the same route and, prompted by his guide, slipped at the same spot. He put down his fore-knees and stayed in that position. Thus, either because I was two stones heavier than Shrinagesh, or because my horse was less skilful than his, or because both these factors contributed in equal measure, I found myself sliding over the high wooden pommel (I wished then it had been higher still), down the inclined plane of the horse's neck and into the water. A large rock behind which I fell saved me from being carried away by the torrent, but I was completely soaked. I worked my way over stones and reached the other side. The Losar men looked grief-stricken by this slight contre-temps, but when Shrinagesh laughed and I joined in the laughter, making light of the whole affair, they felt reassured.

We arrived at Losar at four o'clock and had our first sight of human habitation after five days. The tents were already up, and our camp was surrounded by a large crowd of Spitials. It seemed the whole village population, men, women and children, had come out to greet us and study our way of life. They stood there till late in the evening and filled the air with a fetid smell of human sweat, animal odour and rancid yak butter.

The Spitial is an incredibly dirty creature. He never enjoys the luxury of a bath. We heard on good authority (which I was able to confirm later) that he does not clean himself even after the act of defecation. It is said that he (or she) is washed only twice, once at birth and once after death. In between, he rubs copious quantities of yak butter on his person, to keep out the cold and to counteract the ill-effects of the dry dust-laden winds which are a permanent feature of his land. He rubs the butter into his hair, which is woven into a plait once and for all. The women plait their hair in a large number of fine strands into which are woven vast quantities of yak's hair in order to increase the bulk and length of the natural hair. The strands are spread out over the back and held in place by a string of beads, turquoises or other coloured stones, fastened across the warp. The over-all picture is a broad net of dirty, greasy strands, suspended from the back of the head and extending over the entire back to a point below the hips, and in some cases as far down as the knees.

Monks and nuns do not wear long hair and keep it shaved or shorn very close.

The dress of the men consists of a long loose-fitting coat which I have already described, an undercoat or vest of cotton (or silk if the wearer can afford it) and loose woollen pyjamas. The footwear consists of long boots with soles of leather, and cloth or felt tops reaching up to the knees. The pyjamas are tucked into the boot-tops which are gathered below the knees and fastened with a string to keep the cold out. The woollen rope, coiled round the waist, plays an important part in the attire. It serves to keep the internal organs, especially the stomach, liver and kidneys, warm; it gives support to the spinal column and prevents backaches; and it converts the upper part of the coat into a blouse with a pair of roomy pockets which serve as receptacles for almost anything. Indeed, a Spitial going on a journey will carry in this manner everything he needs on the way. His prayer-box, his prayer-wheel which he will take out every now and again and rotate in a clockwise direction by a gentle swaying movement of his hand, his wooden, copper or silver cup for his tea or water, his tobacco pouch, his bag of *sampa* (parched barley flour), a spare garment, a newly born lamb can all be accommodated in these pouches, and their proximity lends warmth and comfort to his person. The unloading of a well-stocked coat is a fascinating process.

Needless to say that the rope can also be used as a rope, should the occasion arise, though the Spitial is very chary of diverting an essential article of his dress to base uses. Sometimes a woollen sash takes the place of the rope, particularly in the attire of monks. Attached to the cummerbund of cloth or rope may be seen a variety of articles, an iron pipe, a knife, flint, steel tinder box, a metal spoon and a bunch of strange-looking keys.

The dress of the women is similar, except that their coat is always of a dark colour, and they frequently wear a shawl over their shoulders. They usually go bareheaded. Their footwear is similar to that of men.

The Spiti women are very fond of jewellery, and nearly everyone we saw wore several necklaces, ear-rings of elaborate design, bangles and head ornaments.

The necklaces consist of silver discs, glass beads, lumps of coral, amber, or mother-of-pearl, and turquoises, strung together singly or in various combinations. The earrings and pendants, as also the bangles, are made of silver, studded with turquoises or other stones which are imported from Tibet. Similar ornaments manufactured in India are now seen as part of the stock-in-trade of Tibetan pedlars who offer them for sale in Simla and Delhi, with the assurance that they were imported from Tibet.

The *berag* worn by married women is the most striking and the most picturesque ornament; if, indeed, a thing of such magnitude can be called an ornament, though there can be no doubt that it is designed to adorn and embellish. It consists of a large piece of felt or padded cloth, two feet long and nine or ten inches broad, covering the head and hanging down the back over the hair. It is studded with turquoises, silver talismans and glass beads; it is attached to the forehead and to the ears by means of small silver tubes and chains. The *berag*, covering and concealing, as it does, the wearer's hair, gives the impression of an elaborate coiffure, whereas it is more like a headdress worn over the hair. Unmarried girls do not wear the *berag* ; they wear a single turquoise on the forehead, at the point where the parting of the hair begins. That is how we learnt that there were quite a few old maids in Spiti, and we observed that, like old maids elsewhere, they were disposed to become coy and giggly at the approach of strange men. It is possible, of course, that they had heard of the three unmarried men in our party.

Life for the people of Spiti is very hard and connotes

a great deal of scarcity and deprivation. The winter season lasts for nearly six months, and for the greater part of this period even internal communication ceases. The Spitial locks himself up in his house like a hibernating animal, scarcely venturing out of doors even for the purpose of performing the functions of nature. We heard that if an infant dies during the winter months his dead body is buried in the walls of the house, for the ground outside is frozen and covered with snow, and fuel is too scarce to be wasted on cremating a mere infant. For eight months of the year the valley is cut off from the rest of the world. At the end of September the passes that give access to Spiti are snowed up and rendered impassable, and not till the beginning of June can they be traversed with safety. Arable land is scarce, so villages are small and sparsely spaced. A Spiti village is no bigger than a hamlet consisting of twenty or thirty houses, and it lies at a distance of four or five miles from its neighbour. This means that land extending over four miles is scarcely sufficient to support twenty families. There is nothing surprising in this as culturable land lies in a narrow strip, and the total area available is small. The territory of Spiti measures 2931 square miles and the area under cultivation is only 2374 acres. The population is correspondingly small : it was 2422 at the census of 1951.

The Spiti valley was originally a gentle sloping plain, one or two miles wide and extending along the river-bed. The central portion was eroded by river action to a depth varying from twenty to five hundred feet. The process of erosion was in all probability accompanied by the deposit of alluvium dislodged from

the upper reaches of the valley. So the Spiti river now
flows over a broad bed, enclosed by high perpendicular
cliffs, and above these cliffs, on either side, lie narrow
plateaux on which villages and fields are situated. From
the plateaux rise long steep slopes of rubble and debris
ending in vertical walls of rocks and jagged ridges which
look like the fortifications of a gigantic citadel. An
occasional stream irrigates a small portion of the plateau
and makes cultivation possible. Many of the small
streams are lost in the great heaps of rubble and stones
at the edges of the plateaux, and, for lack of moisture,
large patches of land remain barren and unproductive.
Artificial irrigation is rare and not very satisfactory.
There is considerable room for improvement in this
direction and something is being done to encourage and
facilitate the construction of more irrigation channels.

The only crops grown in the valley are barley, buckwheat, green peas, *sarson* (rape-seed) and *tsè tsè* (*Pani-
cum miliaceum*). Owing to the scarcity of grazing and
vegetation, there is very little wild game, and it is
uneconomic to rear animals for meat. So by force of
necessity rather than in obedience to Buddha's teaching
of *ahimsa*,[1] the diet of the people is vegetarian. Let it be
said to their credit, however, that they are not averse
to eating the flesh of a sheep, goat or yak's calf if it dies
a natural or an accidental death.

Their normal fare is frugal and lacking in variety.
The morning and evening meals consist of a soup or
thin porridge, made from parched barley meal. The
meal is stirred into a pot of boiling water and seasoned

---

[1] Non-violence. The believer in *ahimsa* does not inflict pain or
injury on any living being.

with salt.   Vegetables if available — peas, *sarson* leaves
or cabbage — are added and, on special occasions, meat.
Dried peas cooked with butter and spices are eaten with
chappatis of buckwheat flour.   These *chappatis* are made
from dough which has been leavened with a small
quantity of flour kneaded with milk curds.   The *chappatis* are not unpalatable, and are soft and spongy like
pancakes.   The barley meal or *sampa* is also eaten mixed
with butter or tea, or with just plain water.   A bowl of
milk curds, sweetened with coarse country sugar, is
considered a delicacy ;   and whenever we visited a
monastery, or went to the house of an important per-
sonage, we were served with curds in a silver or jade
bowl.   The green leaves and stalks of the *sarson* plant
are cooked as spinach.   Oil extracted from *sarson* seeds
is eaten with *sampa*, and used as a cooking medium and
for lighting lamps.   The refuse of the seeds, left after
pressing, is valuable as food for milch cattle.

An important item of the diet is tea ;   but the tea-
leaves used in most houses are of an inferior quality.
They are bought from Kulu, and are of the coarse
variety which is inexpensive.   The import of tea from
China has almost entirely ceased, and, except in some
monasteries, only the local brand is used.   The leaves
are mixed with water and put on to boil.   Salt and
rancid butter are added and thoroughly stirred into the
concoction.   The final preparation is poured into a
brass or copper tea-pot and served in silver or copper
cups.   In one of the monasteries we were offered tea in
cups of jade encased in bowls of pierced silver.

I had heard a story that the lamas stir their tea with
a human tibia bone, and then blow on it through the

bone. This is supposed to bless the beverage and bring good luck to the drinker. I never witnessed this ritual at any of the monasteries we visited, but probably the ceremony of blessing the tea was performed before the tea-pot was brought into our presence, and this may account for the peculiar taste of the liquid which we felt obliged to drink. Hot tea with butter added to it is a warming drink. Its calorific value is high and, although in Spiti good tea is a luxury enjoyed only by a few, some kind of decoction, bearing a remote resemblance to the real beverage, and possessing at least its heating properties, is drunk by everyone and as often as possible. That is why the Spitial always carries his cup with him in the folds of his coat, so that there should be no distressing interval of waiting between the arrival of the tea and its consumption.

A more efficacious protection against cold is provided by *chhang* (beer made from barley water) and *arak*, the local whisky, which is a kind of distilled spirit not unlike vodka in appearance, odour and taste. Every house has its own still, and everyone brews his own beer. So, in a single village, you may taste thirty different kinds of ale, and enjoy the varying bouquets of as many brands of whisky. As might be expected, the quality differs from house to house — the flavour of some is so strong and nauseating that they cannot be imbibed without doing violence to one's organs and senses, while others are quite palatable and their consumption is a pleasant experience.

The yak plays an important rôle in the economy of the Spiti valley. This quadruped is a big, lumbering, slow-witted but fierce-looking beast, not unlike the

G

bison in appearance. It is difficult to train, and is controlled by a rope passing through a hole in its nose. The yak is used for ploughing, and serves as a mount and a beast of burden. The female provides milk and butter. Yaks are usually black but they frequently have white tails. These are cut off, mounted on silver handles and used as *chowris* [1] in Sikh temples. In Spiti, they are used to drive away evil spirits, and may be seen fixed on houses like flags or weather vanes.

The pure-bred yaks are imported from Tibet and are crossed with cows. The offspring, if male, is called a *dzo* and, if female, a *choru*. Yaks, *dzos* and *chorus* are sturdy animals and feel comfortable only at an altitude of 11,000 feet or more. They have heavy bodies and short strong legs. They move slowly in a field, and have to be coaxed and shouted at when they are put to plough ; but they are very sure-footed, and can run up or down the steepest and narrowest path with confidence. To see one of these monsters come galloping down the hillside is an amazing sight ; one expects to see the whole solid mass of bovine immensity roll over and go hurtling down to the bottom like a boulder dislodged from its seat on the hill-top ; instead of which one sees the beast pursue its devious course along a scarcely visible track and arrive at its destination undamaged and unruffled.

Much of this knowledge I acquired subsequently and piecemeal, during the days that followed, as we toured the Spiti valley, going from village to village, visiting

---

[1] Ornamental brush used for swishing flies, or as a fan. A *chowri* made from yak's tail is one of the insignia of royalty.

monasteries, meeting the people, hearing their griev-
ances and examining ways of ameliorating their lot.
It was then also that I saw the valley in its different
moods, saw how niggardly it is in giving forth its fruit,
and heard of the very sensible way in which the people
of Spiti have found an answer to their economic
problems.

I shall postpone a fuller discussion of these and other
kindred matters to a later stage, and resume the narrative
of our first day in Losar which was also our first day in
Spiti. Losar literally means 'south water', and derives
the name from its position on the river-bank at the
southernmost point of the West Tibetan Empire. From
our camping-ground, situated on an elevation above the
village, we had·a good view of the twenty or thirty
houses standing together in a tidy bunch. A few stray
huts lay irregularly scattered over the slope running
down to the water's edge, as if some untoward and un-
expected impediment had halted them in their precipitate descent from the mountain-top, while the rest of
the village had safely rolled on and come to rest at the
bottom of the valley in a single heap. Losar is 13,380
feet above sea-level and is the highest village in this
region.

We had our tea sitting in the sun, in front of our
tents, while the entire population of Losar watched our
antics, whispering to each other and giggling. Somebody brought a pair of yaks and began to parade them
in front of us. The large ugly brutes hung back,
snorted, shook their heads and strained at the ropes
which held fast their noses, but finally went through
their paces. They walked to and fro, stood to attention

shoulder to shoulder and facing one another, while we photographed them from different points and various angles. We enquired if any *chowris* were available for sale. No, there were no yaks' tails in Losar, not a single one ; but in the next village the headman would certainly be able to satisfy us and procure as many as we wanted. Needless to say we met a similar denial and a similar promise in the next village and the next, and failed to secure a tail from anywhere in Spiti.

After tea we called Turup Dorje, the headman of Losar, and asked him if we could pay a visit to his village. We were conducted through the two or three narrow lanes which gave access to the closely packed bundle of houses. The lanes were surprisingly tidy and the houses had an appearance of having been freshly whitewashed. We saw the interiors of three houses : the first belonged to Turup Dorje, the second to a lama and the third to a peasant of moderate means. The pattern of all three was identical, the difference lay only in the size and number of rooms, and in the quantity of fuel and fodder stocked on the roof-top.

We stood in front of a small opening, scarcely larger than a window at street level. Beyond this opening was darkness. Suddenly Turup Dorje laid hold of Shrinagesh's hand, and, grasping it like a vice, ducked low and disappeared into the darkness, dragging his captive with him. Shrinagesh threw out his free hand just in time to catch mine and pull it after him. I did likewise to the person standing next to me, and so, like a chain of blind animals going underground to burrow, we proceeded to unravel the mysteries of Turup Dorje's house.

The darkness we had entered was a low-ceilinged room full of a heavy dank smell. We had to crouch low and maintain an even tension between the opposing forces operating on the two hands. We learnt afterwards that this was the part where cattle lived and generated some measure of warmth by the accumulated radiation of their body heat. We found ourselves floundering and stumbling up the steps of a narrow, low and tortuous staircase. Then we saw daylight once again, and stepped into the courtyard which is situated on the first floor of the house. On three sides of this courtyard or terrace lay the living-rooms, and on the fourth a low wall overlooking the river and the valley beyond. All the rooms had narrow low doors and were dark and stuffy inside. There was no furniture in the rooms, and a small circular hole in the roof, of which the diameter was not more than six inches, provided the means of ventilation and an egress for smoke. Every room we entered was full of smoke and suffocating fumes. It was apparently cooking time in Losar.

One of the rooms was set apart as a private chapel or prayer room. This was the tidiest and best looked-after part of the house. The floor was clean and freshly washed with mud. The whitewashed walls were decorated with *thankas* or holy pictures painted on silk. A figure of the Lord Buddha in brass rested on a wooden altar in the centre of the room. In front of the altar was a low Chinese lacquer table on which lay arranged a row of carefully polished brass lamps. One of these was burning and cast a dim light in the room. The fuel used was butter made from yak's milk, and the lamp was replenished three times a day. One filling lasted eight hours, and the lamp was kept burning day and night continuously. A thin column of grey smoke was rising from a brass bowl on the altar and the air in the chapel carried the sweet acrid smell of incense. A *dorje* bell beside the incense burner represented the power of thunder.

In the house of the lama the chapel contained a beautiful Trimurti. The figure was carved out of wood and covered with thick gold leaf. The *thankas* in this chapel were better and more skilfully executed. We asked the price of one of them, not because we wanted to buy it, but because we wished to translate its artistic and religious value into material terms. The lama said it would cost not less than two hundred rupees in Tibet. In the peasant's house the materials of worship were cheap and unattractive, though I have no doubt that they inspired as much religious fervour in the peasant's breast as the more glittering paraphernalia of the monk's chapel.

Over the flat roof lay a thick pile of fuel and fodder. There are scarcely any trees in Spiti, except a few willows planted in some well-watered patches near villages, and for fuel the people use dried dwarf juniper bushes. These are sought out and brought home by the women-folk from far distances. Throughout the summer months the women of Spiti may be seen carrying loads of these bushes, and waddling along slowly with their backs bent double under the weight. Incidentally it is the women who do most of the work in this happy land.

Our activities continued to excite a lively interest in the people, and a large crowd of spectators followed us

from house to house, and closed in upon us as we
arrived back in camp. They could not be persuaded
to return home in spite of the lateness of the hour. So
we turned on the radio, and when they heard strains of
music and human voices issuing from the small magic
box, they laughed unrestrainedly. We turned the knob
round and the music faded out. We shook our heads,
made helpless gestures and looked up at the sky in
despair. We told them that if they went home there
would be more music in the morning.

Finally they all went home and we sat down to a
plentiful meal of hotch-potch and tinned mango slices.

The next morning I was lying in bed, relaxing, in
anticipation of a late start and a light day, when I heard
a familiar babble of voices outside my tent. My first
thought in the state of semi-somnolence was that a
movie crowd was practising the 'rhubarb, rhubarb'
technique, but almost at once I knew the truth and felt
reassured. Bhavnani's cry of distress rang out above the
mumbled chorus of crowd noises :

'Shrinagesh, Shrinagesh.'

'Yes ?'

'Is there any place where one can sit down quietly ?'

'Sit down ?'

'Yes, you know. The whole place is swarming with
people.'

'Yes, of course. Go back the way you came yesterday, for about a mile.'

'Good lord, a mile!'

He was away for three-quarters of an hour, and when
he returned we were dressed and ready for breakfast.

Everything moved slowly. We had planned to

leave at half-past nine but at ten not a single mule had been loaded. The mule-men had been late in coming, the horses had wandered far to look for more succulent morsels of green fodder, and the servants seemed in no hurry to pack up. I told Shrinagesh that the graces and the hospitality of the ladies wearing turquoise solitaires had undoubtedly induced a state of porcine sloth in our men. There was very little we could do, except shout and swear at the men, to work off the charm. Finally we were able to make a start at eleven o'clock.

We crossed the Spiti river on horseback. Each horse was led by two men and we had to bend back our legs to keep our feet from getting wet. There was a horrible moment in midstream. The current was fast, and our mounts hesitated. They seemed disinclined to proceed farther, but the men led them on, and soon we were scrambling up the opposite bank.

The route to Kioto, our next halting-place (nine miles from Losar) lay over the plateau on the left bank. The villagers had organized a community project and worked on the path, making it broader and smoother. It seemed quite like a return to the civilized world. There was even more of the civilized world in the crowd of men who followed us, shouting and gesticulating. We learnt that this demonstration was directed not against us, but against the people of the next village, Hansa, with whom the Losarians had a dispute concerning grazing rights. The Hansaites were not yet in sight, but they would be waiting on the boundary of their village, and it was important to work up enough steam beforehand, so that the Hansaites should see just how intensely the Losarians felt about the matter. A little

later, when the men of Hansa faced the men of Losar
with dark and menacing looks, I thought I was back in
the days when, as a young district officer, I had settled
village disputes about boundaries, canal water turns and
rights of way. But the Spitial is not so hot-blooded
and violent as the Punjab peasant; he is, on the whole,
lazy, good-natured and cheerful, and cannot be bothered
with violence in any form. So it was that both parties
were content to leave the decision in our hands, once
they realized that we had taken note of the carefully
staged hostilities. The Losarians returned home and
the Hansaites fell into a state of beatific tranquillity, and
followed us all the way to Hansa without speaking a
word.

At Hansa we visited the local monastery, and the
monks served us with milk curds, which I found quite
palatable. Outside the monastery, a scene of high
emotional tension was enacted for our benefit. All the
women of the village collected, and, making a ring
round Shrinagesh, set up a mighty chorus of wailing.
They shed tears of anguish and prostrated themselves
on the ground. Some of them clutched at Shrinagesh's
feet, others grovelled in the dust before Bachittar Singh,
while the rest merely sat and rubbed their shawls over
their eyes ostentatiously. Our interpreter informed us
that this display of sorrow and distress was related to the
dispute about grazing rights. Shrinagesh told them
that he had taken the statements of both parties, and
had directed the *patwari* to make accurate measurements
of the pasture grounds. When he came back after a
few days, on his return journey, he would examine all
the material and pronounce judgment.

Ultimately he asked me to intervene and I suggested an amicable settlement. The two headmen were persuaded to sign a deed of compromise. One or two villagers voiced their dissent, and Shrinagesh asked me if the terms of the deed could be legally enforced.

'Only as between the actual signatories, if the matter goes before a court ; but let them think they are all bound by the undertaking given by their headmen. Of course, we could make everybody place his thumb-impression on the document, but even if one or two refused you would be exactly where you are now, because they are all proprietors and co-sharers in the village land.'

So the matter was left at that.

Two years later I was inspecting the court of the Subordinate Judge at Kulu. While turning over the pages of one of the files, I thought I heard the echo of names I had heard before. I read through the case. It was a suit brought by the proprietors of village Hansa against the proprietors of village Losar. The plaint was based on the allegations that (a) the men of Hansa had never entered into any compromise with the men of Losar, (b) the headman who appeared to have signed the deed had no authority to contract on their behalf, and (c) the headman's signature had been obtained by misrepresentation, fraud and undue influence (the name of the guilty party was mercifully not mentioned), and so the compromise did not bind even the headman.

I do not know what the decision of the court was, but it was interesting to read that I had practised misrepresentation, fraud and undue influence. I did not reveal to the Subordinate Judge what part I had played in the dispute. That might have influenced his judgment !

The portion of the valley between Hansa and Kioto reminded me of pictures of the Grand Canyon in America — lofty precipices, rising perpendicularly for thousands of feet and ending in jagged crags, in colours of pale pink, bright red and purple. The camping ground at Kioto is set in ideal surroundings, in a clearing above the village. Behind us lay steep slopes of rubble, on three sides of us gigantic boulders of incredible dimensions stood sentinel, enclosing us like the fortifications of a rugged castle. In front of us the plateau sloped gently down to the village for a hundred yards. Beyond was a mountain wall with snow-covered peaks and the corner of a glacier peeping inquisitively from behind a rocky prominence. Below the glacier gaped the dark mouth of a huge grotto.

In these surroundings one could spend days of inactivity, watching the clouds form, disperse and reform, and the colours change. I had seen Nicholas Roericks's paintings of the inner Himalayas, and had taken them for idealized conceptions of the great master, bearing scarcely more than a remote resemblance to reality. But in Spiti I saw purple and blue mountains outlined sharply against yellow and red skies. The transparency of rarefied air and the dispersion of ultra-violet rays at high altitudes do strange things to the light of the sun, and as the splendour of the day increases to the fullness of noon, and then gradually fades away into the twilight of evening, the colouring of the landscape alters from hour to hour, ranging through all the bands of the spectrum.

Our somewhat unrestrained outburst at Losar bore fruit at Kioto. The men had taken to heart our remarks about their incompetence and inattention, and began pulling up the tent-pegs at half-past four in the morning. I scrambled out of bed and dressed quickly. At six o'clock we sat down to breakfast; and at seven we were on our way.

The path ran quickly down to the Spiti river, came to a point where the Takling stream joins it, and went up a steep bank on the opposite side. We forded the stream on horseback without any difficulty. The river-banks at this place assume the most fantastic shapes. The soil is a closely packed mixture of sand, pebbles and small stones. The softer portions have been washed away by rain-water and river action, leaving strange spires and steeples growing out of the banks like some malignant growth, developed by a torpid monster. Some of these prominences, by their size, shape and symmetry, resembled the towers of skilfully fashioned temples, others looked like the voluted columns of a Gothic cathedral, and from a distance the pattern of pebbles with which they were studded gave the appearance of carvings in stone.

At the bottom of the valley we paused for a moment and cast our eyes around us. The bleak barren banks rising to a height of more than five hundred feet, the towers and spires massed above us, and the yellow Spiti torrent, roaring below, were like the remains of a dead world which is being disinterred after the lapse of a thousand years. It was an eerie sight and induced a vague feeling of discomfort.

After a stiff climb we reached the top of the Lagudarsi

pass — a series of pleasant meadows of which the largest was nearly half a mile long and over two hundred yards broad. It was almost as flat as a football field and Shrinagesh thought it would make an excellent landing-ground for planes if a few boulders were removed and a small rivulet which went meandering slowly over its surface were diverted or drained off. 'We should be able to come here in winter and ski down these gorgeous slopes, then,' I said.

A few miles farther lay the magnificent Kibar gorge. The river, flowing on a bed of shingle, is enclosed by high vertical walls of limestone, with multi-coloured strata arranged horizontally. The flow of the stream at this place was gentle, and behind a corner was an inviting pool, in which we bathed. A steep path, running up the face of the rocky escarpment, brought us to Kibar village. The women, as is usual in Spiti, were working in the fields while the men stayed at home and gossiped. On this occasion they came out to greet us and help our men in setting up camp.

The Spitial's answer to his economic problems is the law of primogeniture, combined with what can only be described as enforced celibacy for the younger sons. The wealth of the valley is extremely limited, for the total area of land available for agriculture is small, and the yield is scarcely enough to feed the whole population. There is undoubtedly scope for improvement, and if modern methods of crop cultivation were applied and means of irrigation improved, the produce could be doubled. For the present, however, and for hundreds of years in the past, the valley, in its isolation, has had to subsist on whatever scanty means the niggardly soil provides. An increase in population would have proved disastrous, and, to ward off the danger of slow starvation, the Spitial evolved a very sensible scheme of keeping constant the number of mouths which must be fed. The eldest son in each family inherits the land and property of his father and the family residence ; the younger sons take to the monkish profession and remain celibate. So each succeeding generation consists of the same number of families as the previous one, and the population does not increase.

As soon as the eldest son marries, he enters upon his office as head of the family and takes over the ancestral home and lands. His father retires and goes to live in a small house which is called *khang chung* (small house). The *khang chung* is the permanent home of super-annuated parents, and devolves upon each successive father on his retirement. The grandfather, if alive, has a house of his own, and this too descends from one grandfather to another. The younger sons are sent off to the monastery in their childhood and are trained to become lamas. The monasteries are endowed with *jagirs* or revenue-free lands and derive some income from *pun* or offerings made in cash and in kind. Thus the family estate remains undivided, and the entire family is adequately provided for. If the head of the family dies without leaving a son, or if the surviving son is too young to assume the duties of a paterfamilias, the younger brother of the deceased leaves the monastery and takes over the family estate. He also takes over the family, for he has the right to treat his brother's widow as his own wife. She cannot object ; indeed, so well is this right recognized that no marriage ceremony is necessary. All that the ex-lama has to do is to pay a forfeit in the form of a fee to the monastery on leaving.

If the head of a family has only daughters and no male issue, he sometimes adopts a son-in-law as his heir. But if he has a younger brother in a monastery, there is sure to be an objection from the brother's side, and he will claim the right to come home and beget a son by his sister-in-law. 'You, dear brother,' he will say, 'have failed to raise a son by this worthy woman ; now, let me try.' If he succeeds, the elder brother is relegated to the *khang chung*, the small house reserved for ousted and superannuated proprietors. The younger brother may even take a new wife in order to raise a family.

It is not really so complicated as it sounds. The aim of the Spitial is to keep the population constant, and his customary law is designed to achieve this aim. Only one man in each generation is allowed to marry, and this restriction ensures that there will be no multiplication of families. On the other hand, provision is made for the continuation of the family from generation to generation by engrafting a number of exceptions on the general rule of primogeniture, lest a too rigid observance of its terms result in the extinction of the family. The system gives rise to some unusual results. By far the largest number of marriages in Spiti are monogamous, but polygamy and polyandry are also practised and co-exist with celibacy on a large scale. This, however, is a matter of small importance, and the Spitial finds nothing anomalous or strange in his social customs or his law of inheritance.

An interesting consequence of this law is that the

domestic male is a rare animal and is, for this reason, pampered by the women-folk. All the housework and all the work in the fields, with the exception of plough-ing, is done by women. Once a year, in autumn, the men lead their yak teams out to the fields, and with much shouting and screaming turn the soil over to make it ready for sowing. They lay on a dressing of manure and then retire to the warmth and smoky com-fort of their homes, where they lie back like drones, and like drones endeavour to perform their real function in life — the begetting of children. In their moments of leisure, which are plentiful, they renew their vigour and maintain their interest in life by consuming large quanti-ties of *chhang* and *arak*. It may be a dull kind of life by some standards, but I never saw any signs of an urge to do something constructive or creative. The people looked happy and cheerful, and had the appearance of contentment and inner peace which I have come to associate with those who follow the teaching of Buddha.

Kibar is one of the larger villages in the valley, and has a correspondingly greater area of arable land. On three sides of our camping-ground lay green fields, in which barley and green-pea plants moved in slow rhythmic waves in the afternoon breeze. I took a walk through the fields and saw a whole network of water channels. The nature of the surrounding landscape makes it easy to bring the waters of the snow-streams down to the village, and irrigate a large area of land. Kibar is thus a prosperous village by Spiti standards.

I think I must have taken a larger helping of hotch-potch than was necessary to supply the wants of my body, for I woke up in the middle of the night with a feeling that I was being suffocated. I came out of my tent and saw that a gentle drizzle was floating in the air — one could scarcely say it was falling from above. The air was cool and its touch gave a pleasing sensation to the skin. There was complete quiet all about me, and the camp was enveloped in the light mist of rain-drops. I walked around the camping-ground for a few moments and felt better. I remembered that Kibar is situated at an altitude of 13,000 feet, and took a count of my pulse-beats. Sixty-four to the minute. At Karcha, after a strenuous day over the Shigri glacier, it had been ninety-two. That set me thinking. Wasn't my heart taking things a bit too easy ? Wasn't the normal pulse-rate for a healthy man more than sixty-four, and nearer ninety-two ? I went back to my tent, and sitting down, arranged my legs in the lotus pose. After a moment's concentration I started the *ujjayi pranayam* — the simple deep breathing exercise in the yoga way, taking care not to make too much noise. After five minutes the feeling of suffocation disappeared and I slept soundly till the morning.

We had fresh yak's milk with porridge for breakfast. I have heard people complain of an unpleasant odour which they find in yak's milk. For myself I always found it palatable and very sustaining. In colour it is a deeper yellow than cow's milk ; it is thick and very rich in cream and butter. Too much of it will lie heavy on the stomach, but, taken in moderation, it is excellent fare, and whenever it was available I took it in preference to the canned stuff we had brought with us.

After breakfast we paid a short visit to the local

H

monastery. It was an unimpressive, mean-looking cottage without dignity or character, and except for one or two *thankas*, we saw nothing of interest. The lama in charge of the monastery told us that the *thankas* were worth about Rs.200 each. I had a suspicion that he was offering them for sale ; but since Shrinagesh had asked us not to buy anything from the *gompas*,[1] as that might savour of obtaining things by undue official influence, we smiled approval and took our leave.

Two miles from Kibar we left the main route to Gaza and took a narrow track to the Kye *gompa*, where we arrived at a quarter past nine. Information of our visit had been sent in advance, and the monks had come down to receive us at the bottom of the conical hill on which the monastery is perched. We heard the strains of a slow-moving sorrowful chant, played on what we afterwards saw were trumpets of a monstrous size. The older monks were dressed in ceremonial costumes, consisting of a red robe and a scarlet head-dress which is something between a Dutch cap and a cardinal's hat. Some of the monks carried bells, incense burners and other articles of worship. The head lama garlanded each of us in turn with a piece of muslin, about a yard in length. The lamas mumbled a prayer, blessed us, and then led the way to the monastery in a slow, solemn procession.

The monastery consisted of a large irregular heap of low rooms and narrow corridors, inter-connected by dark passages and tortuous staircases, built on what must be the crater of a now extinct volcano. We entered the main courtyard and had our first sight of the strange

[1] *Gompa* — Buddhist monastery.

orchestra whose music had greeted us at the bottom of the hill. There were two trumpets, two or three smaller brass instruments with squeaky notes, a pair of large cymbals, a big deep-voiced drum and a pair of smaller ones. The trumpet, which is eight or nine feet in length, rests on a triangular block of wood, placed on the ground, and rises in a long sweeping curve to the mouth of the player who remains standing and blows into it with all his might. The sound produced by this magnificent instrument is surprisingly feeble. It was distressing to see that almost the entire energy generated by the trumpeter's red face and distended cheeks was being dissipated in such an unproductive way.

On one side of the courtyard was a short flight of stone steps leading up to a door which gave access to the inner rooms. A stuffed ibex and barhal, nailed above the doorway, gave distinction and dignity to the entrance. The skin of both animals had given way at several places, and straw stuffing showed through the holes. A thick covering of dust concealed almost completely the colour and markings of the beautiful coats, but the shape and the horns, particularly in the case of the ibex, still bore evidence of the nobility and poise of the animals. We passed through the door and entered a small room which was decorated like a private chapel with a small altar-piece at the far end, on which rested figures of Buddha and Tara Devi. A stuffed snow leopard, in the rampant attitude, stood snarling at a panther from the plains, overflowing with straw and rags. Dust and neglect lay heavily over the two animals.

We were led to a room upstairs which was furnished with Chinese carpets and had a raised dais at one end.

Shrinagesh and I were given seats arranged behind a low table on this dais. These were the seats of honour — the greater the dignity of the guest, the higher the seat offered to him. In the absence of a dais, the seat of honour is raised by placing one or two cushions on the ground. The others sat close to us, but at another table below the dais.

We were served with milk curds and sugar in silver bowls and *sampa* with yak butter in small silver saucers. The spoons were exquisitely fashioned, and had carved handles. Then came tea, salted and buttered and piping hot, in a beautifully carved teapot of copper and silver. It was poured out into silver cups with a casing of wood to facilitate handling, and had a foul taste. We ate the curds and *sampa* and drank the tea. We were waited upon by a fair-complexioned chubby lad of ten who had an extremely handsome face and dark soulful eyes that gave him a precocious look. This beautiful Ganymede was neatly dressed, and his movements had the correctness and dexterity achieved by natural grace and assiduous practice.

The monastery had a *jagir* (grant of revenue-free land) attached to it and the monks were anxious that this benefit should continue after the new settlement.[1] The head lama laid a muslin piece on the table in front

[1] Settlement operations in respect of any given area (usually a district) are undertaken at intervals of thirty or forty years. The entire area is re-measured carefully, and a new and complete 'record of rights', giving the name of each proprietor and the area held by him, is prepared afresh. The total produce of each village and the revenue payable to the State are reassessed. Settlement operations are a very important feature of land administration in India and are conducted by the Settlement Officer, who in this case was Bachittar Singh, under the supervision of the Commissioner.

of Shrinagesh, and stood with folded hands. Next to him stood the spokesman of the lamas, a tall learned-looking monk with spectacles. The expression on his face was sad, and he appeared to be consumed with pity for the world around him. He spoke in slow, deliberate accents, and paused after each sentence, so that the interpreter should have time to convey to us a complete translation of his pleas and arguments. Shrinagesh said that if the monastery claimed a renewal of the *jagir*, the monks must do something for the people. They should (*a*) make a proper path, six feet wide, to the monastery and (*b*) supply two teachers to the local school in the village.

The first condition was promptly accepted, but the sorrowful monk demurred with regard to the second, and submitted that the cost of supplying two teachers would be ruinous. The lamas were willing to open a school in the monastery, and act as teachers ; but there should be no interference by the State in the management of the school. This proposal was wholly unacceptable to Shrinagesh, who withdrew his hands from the table to indicate that he was not prepared to pick up the muslin piece. The monks deliberated among themselves for a few minutes, and cast uneasy glances at Shrinagesh, sitting adamant, with his hands folded ostentatiously behind him. Finally they agreed. The muslin piece was picked up, and the monks voiced their gratitude in a mumble of happy approval. I am glad to record that the *jagir* attached to the Kye monastery was renewed, and will continue to subsist. The monks, too, are carrying out their part of the bargain, and there is now a well-established primary school at Kye.

On arriving at the monastery we had expressed a
desire to see a performance of the religious ballet by
masked dancers.

'How long would you like the dance to last?' the
head lama had asked.

'How long does it usually take?'

'The complete ballet takes twenty-six hours at the
very least, but we can stage one item which will not
last more than two or three hours.'

We said half an hour would give us satisfaction in
ample measure. While the performers were dressing
for the ballet, we were conducted through the main
rooms of the monastery, and shown the artistic treasures,
of which the monks were justly proud.

The monasteries are all built on a similar pattern.
The chapel and the assembly rooms form the central
core, round which lies a veritable labyrinth of cells,
corridors and passages. In the cells the monks live,
cook their food and keep all their belongings, books,
clothes, utensils, etc. Each family in the neighbour-
hood has its own cell which is occupied by the entire
hierarchy of younger brothers, uncles and nephews.
They come to the family cell as and when they renounce
their secular life and take up the monkish profession.
In the passages and in the corners of the main prayer-
room stand huge prayer-drums of a size greater than
beer-casks, beautifully painted and lacquered and inscribed with sacred texts. As you pass you set them
gyrating in the clockwise direction. So well are they
mounted and balanced on their bearings that they
turn round at the gentlest touch and say your prayers
for you.

In one of the rooms there were several prayer-drums of a smaller size, arranged in the form of a square, occupying the centre of the room. There were one hundred and eight of them and we went round the narrow passage between the wall and the rows of prayer-drums, touching each one in turn. In front of a large prayer-drum in another room lay an open wooden box like a low manger, obviously intended for a penitent monk to sit in isolation and purge himself of his sins.

The walls of the central chapel, where we had tea, were beautifully painted with scenes from the *Jatakas* and from Buddha's life. There were quaintly drawn figures of demons, gods and goddesses. There were several examples of the Wheel of Life, or, as it is more correctly called, the Wheel of Becoming. There was a painting representing the temptation of Buddha at which I gazed for a long time, fascinated by its beauty and craftsmanship. The Lord Buddha sat under the sacred tree with his legs crossed in the lotus pose, and a look of serene tranquillity on his face. A nude female of ravishing loveliness clung to him in a lascivious embrace — her arms coiled round his neck, her legs circling his waist, her warm passionate cheek touching his, and her partly open, rapturous lips seeking his cold passionless mouth, her whole body lifted and poised in anticipation of the erotic climax. The artist who painted this picture — probably a visiting lama from Tibet — must have seen a woman in the act of love, watched the ecstatic movements of her limbs and studied the lines of pain and pleasure on her face as she approached her orgasm. He was a most careful and accurate observer, but I very much doubt if he was a celibate monk, condemned to a life of frigid abstinence. There were paintings of the goddess Taradèvi and of dark-skinned Durga, standing naked and menacing, with legs parted and arms raised, ready to strike and destroy. *Thanka* after *thanka* was unfurled and displayed before us. A month would not have sufficed for even a partial appreciation of so much art ; all we could do was to look at each picture for a few moments and give an occasional gasp.

The pictures were all drawn with great skill and artistry, and painted with brilliant, well-balanced colours. For colour the artist uses a paste which is prepared by rubbing natural stones of different colours on a wet surface. Colours made in this way are fadeless and retain their brilliance almost indefinitely. The artists are, in nearly all cases, hired from Tibet and commissioned to do the work. In some cases a visiting lama may leave behind a souvenir of his visit in the form of a painting or a manuscript. The *thankas* in the Kye monastery were without question the best I have ever seen. They were, for the most part, brought from Lhasa by the local monks who went there for study or pilgrimage. Silver, jade and amber cups, *dorje* bells, incense burners fashioned like gargoyles, brass and jade figurines were neatly arranged on the altar. Behind the altar, which ran along the entire wall, was a bookcase containing the hundred-and-eight sacred books.

It was fascinating to see and examine these masterpieces of religious art, but the air in the room was heavy with smoke and the pungent odour of incense. We were glad when word was brought that the masked dancers were ready to begin the ballet. We went down to the courtyard and took our seats on the stone steps.

The music began with a slow-moving chant which had a distinct rhythm, emphasized by the beats of the large drum and the clanging of cymbals. The melody was mournful and ranged over not more than two or three notes, sounded by the brass instruments. Two masked dancers made their appearance. They wore gowns of rich Chinese silk and cloaks of gold brocade. The gowns were embroidered with brightly coloured figures of demons and dragons. The heads and faces of the dancers were hidden beneath huge masks, representing hideous monsters. In their hands they carried lacquered batons. They began to move with slow, even steps. Soon they were joined by two more dancers, dressed even more gorgeously and wearing even more grotesque masks. They were armed with queer-looking weapons. Two more arrived, and as the group of six moved round the courtyard the *tempo* of the music became gradually faster and livelier. The dancers represented the demons or forces of evil, making merry and extending their sway over the world. The triumph of the demons and the tyranny to which they subjected mankind were symbolized by a variety of gestures and dance movements which were repeated so many times and over so long a period that we began to tire of the performance.

At last the gods made their entry on the scene, to join combat with the demons. They were similarly dressed and masked, for the legend says that when the gods heard of the cruelties and impious barbarities practised by the powers of darkness, they left their heavenly abode and came down to earth disguised as monsters and strange beasts. They were thus able to meet their adversaries on their own ground, and, by a simple trick, mislead them and completely destroy them. The engagement of the gods and demons was the high light of the ballet ; and as the battle progressed the music swelled and advanced to a quicker measure, till the dancers were gyrating like spinning tops and weaving an intricate pattern of fast-moving figures. As they moved they crossed their weapons and struck at each other, synchronizing the strokes with the beats of the drum, which became more and more rapid. The dancers were panting with exertion and excitement. The performance had lasted an hour-and-a-half, and some parts of it were quite tiresome ; but the final climax was well worth waiting for.

Suddenly I thought of the men who were wearing these rich robes imported from China at great cost. Each one of them must cost hundreds of rupees. I had never before seen such gorgeous silks and brocades of such extraordinary brilliance and design. Any one of these costumes would make a sensation at a fancy dress ball in the West. In Kye the men who wore them and staged this glorious pageant were dirty, grimy monks who had never known the luxury of a bath and who literally stank to high heaven. Most of them were illiterate and without any knowledge of what the teaching of Buddha meant. Once a day they met in the assembly hall of the monastery and repeated their prayers, parrot-like, after the abbot ; some of them worked on the lands attached to the monastery, and looked after the cattle. For the rest, they swivelled round their prayer-wheels while their thoughts dwelt on the profanities of this world ; they cooked their food and drank their liquor, and lived in their cells like pigs in a sty. Such, at any rate, were by far the majority of men who joined the monasteries, not because they felt a religious urge or a yearning to achieve *Nirvana,* but because they were driven away from their homes by the pressure of economic forces and social laws. Some there must be who study the scriptures and try to understand their meaning ; or who practise austerities to mortify the flesh, so that they may free their minds from the bondage of *maya* [1] and strive for salvation ; but of these the number is pitiably small.

The dance was coming to an end. One by one the demons accepted defeat, and retired to the darkness of the nether regions. The gods, left in sole possession of the universe, made their exit in a slow and stately triumphal march.

We bade farewell to the monks and, leaving the monastery, took the steep and winding path which goes down to Kye village, at the bottom of the conical hill. My thoughts dwelt on the monastic system of Spiti, and I could not help feeling that life in the *gompas* is idle, slovenly, wasteful and unproductive of any material or spiritual wealth. There is no attempt at order, sacrifice or service. The lamas have, no doubt, some duties assigned to them — they till the land attached to the *gompas,* rear and look after the cattle ; once a day they chant their prayers in the assembly hall, once a year they hold a festival, lasting a week or ten days, at which they entertain the neighbouring villages

[1] Illusion that is the material world.

with masked dances, and give free food to those who take the trouble to come. But for the rest they gossip, eat, drink and sleep. I have no doubt they also fornicate, and indulge in other forms of sexual excesses ; for though women are forbidden to enter the monastery, the monks are free to go out. In the Kye *gompa* we had seen several young lads, besides the beautiful Ganymede I have spoken of, who were on probation as novices, and I suspect that the older lamas do not hesitate to use them for the satisfaction of their libidinous appetite.

The monks make no effort to keep the monastery clean and tidy, and over everything — *thankas*, walls, floors, silk hangings, holy books and furniture — there lies a thick coating of dust. They do no constructive work ; there is no attempt to make the slightest contribution towards religious or philosophical thought ; by far the large majority of them do not even understand the teaching of Buddha. One by one they have forgotten the precepts of the Master and abandoned the path pointed out by him.

The discourses of Buddha, as they have come down to us, refer to an elaborate monastic system, but in the early days there were no priests, for there was no god to worship and no ceremonies to perform. There was no hierarchy and no central authority to check healthy development in new directions, or to impose rigid rules of thought and conduct. Buddha laid down a body of disciplinary rules for the guidance of those who were seeking knowledge. Buddhism at that time was not merely a rule of life for ascetics, but also a moral code and a system for laymen, applicable in all duties of life. There were five moral rules binding on all lay people :

refrain from killing; from taking what is not given; from wrongful indulgence in the passions; from lying; and from intoxicants. For the monk there were, in addition, special rules :

'He eats at the right time, does not see displays of dancing and music, does not use garlands, scents and ornaments, or a high bed. He does not take gold and silver and certain kinds of food, or accept property in slaves, animals or land. He does not act as go-between, or take part in buying and selling, and the dishonest practices connected therewith.'

At the advanced stage the monk was enjoined to avoid 'the acts and practices of which the brahmins were specially accused, such as the interpretation of signs on the body, portents, dreams, marks made by rats, the performance of various sacrifices, and magical ceremonies, the interpretation of lucky marks on things, persons and animals, prophesying victory to an army, foretelling astronomical events, famines, epidemics, lucky days and the use of spells.' What a far cry this is from what we saw in Spiti!

With the development and rise of the Mahayana school of Buddhism in India, the monastic order acquired magic and divination as part of its religious practices. When Buddhism (of the Mahayana school) was introduced into Tibet in the 7th century, it was further influenced by the *Bon* religion which prevailed locally. The order developed a hierarchic organization, the head lama assumed gradually increasing power, till, in the 14th century, he asserted temporal sovereignty. In the following century the Dalai Lama was established as the religious and temporal head of Tibet with his headquarters at Lhasa. The Mahayana sect believes that every individual can become a Buddha, and one who has vowed to attain Buddhahood is repeatedly born to show others the way of wisdom. The Dalai Lama is the reincarnation of the great *bodhisattva*, Avalokitesh-vara, the ancestor of the Tibetans. After the death of each Dalai Lama a country-wide search is started and a highly developed system of magic is employed to deter-mine his successor.

It has been said that the Lamaism of Tibet is 'perhaps the most utterly corrupt form of the religion of Gautam'. The lamas of Spiti owe their allegiance and loyalty to the Dalai Lama and the *gompas* are subject to his control. Lhasa is the emotional focus of Spiti as well as of Lahoul. The Lamaism of Spiti has been greatly influenced and contaminated by local legends and superstitions and the indigenous demonology of the mountains. There are many benevolent spirits and malevolent demons who are supposed to dwell in trees, rocks and hill-tops, and before whom the Spiti Buddhists sacrifice sheep and goats. They believe greatly in witches, sorcerers and the evil eye.

This form of debased Lamaism keeps its hold on the people by old association and by methods of terrorism. The simple, unsophisticated people of Spiti cling to their old beliefs, and pursue an unaltered course of con-duct, any digression from which is threatened with severe pains and penalties. There is a ban on the plant-ing of trees and the opening of new sources of irrigation. In some cases even old water-ducts are neglected. We saw several untapped or wasted springs, streams which run into debris slopes and are lost, land which would

give a good yield if properly tilled and irrigated. The
monastic system finds support in the law of primogeniture, for the eldest son keeps control of the family
lands, and if his younger brothers did not take up the
monkish profession he would have to give them an
equal share. When a lama starts for Lhasa to take his
degree the head of his family gives him something towards the expenses of his journey.

The social and economic life of the valley is inextric-
ably tied up with the monastic order. On the death of
an individual, funeral offerings are made to the monastery on behalf of the family, and these consist of money,
clothes, pots and pans, grain, butter, etc. At each
harvest parties of five or six monks go out to beg alms
all over Spiti. They go round from house to house in
full dress, and, standing in a row, chant verses, the
burden of which is — 'We are men who have given up
the world ; give us in charity the means of life ; by
so doing you will please God, whose servants we are'.
The receipts, we were told, are considerable, as every
house gives something to every party.

A break-up of the monastic order would necessarily
entail the break-up of the social and economic order,
for with the emptying of the monasteries the authority
of the eldest son would go, the population would increase ; but with improved means of cultivation the
valley would be able to support a much larger number
of inhabitants. Steps are being taken by the government to encourage the planting of more trees and the
construction of more water-channels. There is, however, no attempt to change the social or monastic order,
and in the meantime the eldest sons plough the land

and then loaf through the pleasant summer till their industrious women-folk have brought the crop to maturity and cut it. The monks, too, then emerge from their cells and enjoy the life of the village. What a waste of man-power this entails!

The religious life of the people is restricted and stunted by fears and superstitions, but the Spitials are not without a means of escape and salvation. *Om manipadme hum*, the saying of Lord Buddha, is the open sesame to every door, sacred or profane. It is the thread of Ariadne which takes them safely through the maze of doubts and phobias. Its repetition, conscious or unconscious, vocal, manual or mechanical is the panacea for all ills, and it washes away all sins. The Spitial utters the words as he wakes up in the morning; he repeats the *mantra* when he is telling over his beads; he inscribes them on a copper or silver wheel, and spins the sacred words round and round in the clockwise direction; he paints them on a small or a large drum, and makes the drum rotate on its axis; he carves them on a stone slab and leaves them to mingle with the chorus of a thousand other slabs; he prints them on a piece of cloth and lets it flutter in the wind, so that with every flutter his prayer goes up to heaven. Sometimes he fastens the prayer-drum to a mechanical contrivance such as a water- or a wind-mill, and lets the forces of nature manufacture piety for his benefit. It is an arm-chair type of religion, but it also possesses all the discomforts and qualities of a static state. The man who stays in his arm-chair too long feels stiff and uncomfortable in odd places, and finds it difficult to get up and move forward.

We were at the bottom of the hill, and in front of us the Kye village lay huddled together in a small bunch of houses. We skirted the village and went down to the bed of the Spiti river, which at this place is flat and about two hundred yards in width. The main current had moved over towards the right bank, leaving an expanse of level ground in which stretches of grass and sand alternated with stony patches and little rivulets. After two miles of this pleasant terrain the path hugged the left bank and we had to walk over stones and boulders. Shrinagesh and I sat down to eat our midday meal. Cold *parathas*, cold scrambled eggs and a wing of cold roast pigeon made a satisfying meal, and there was really no reason why my thoughts should have run amok and awakened memories of *à la carte* dinners eaten in Paris many years ago and the taste of red wine. I remembered in particular a bottle of *rosé* I had consumed at a small village restaurant in the province of Burgundy on a bright sunny day in August, and told Shrinagesh about it.

As we approached Gaza we crossed a small stream, on the banks of which grew a profusion of wild rose bushes. They were all in bloom, and one bush in particular had so many flowers on it that the leaves were completely hidden, and from a distance all we saw was a large splash of bright pinkness staring out of a confused patchwork of spotted greenery.

We were to halt at Gaza for two nights and a day. The prospect of a late morning and a day of complete rest induced in me a feeling of sensual ease. I pondered over the last few days of the trek and the tension caused by the continuous haste it had involved. Sometimes

I

one had felt that one was driven on by a relentless hand, regardless of aim or purpose. Wake up at four or even earlier each morning, dress hurriedly in a semi-recumbent position, roll up the bedding and pack up the box, counting over the various articles lest anything should be forgotten or left behind. Breakfast, and a long hour of tiresome waiting while the mules were loaded ; and then, march on. Repeat the drill the next day and every day. No time to stop on the way and drink in the beauty of the landscape in full measure.

The past few days had been like a long night in which dreams of loveliness alternated with horrible nightmares. Caves of ice and caverns of glistening snow; tall, rocky massifs that change their colours from hour to hour, and visions of blue skies with clouds that appear and disappear, intermingled with the shock of murderous streams roaring to swallow man and beast; steep, treacherous moraines and precipices waiting for the unwary traveller to take a false step; and narrow, slippery ledges overhanging torrents which writhed and groaned a thousand feet below. In the heart of the Himalayas Nature is not passive. Her beauty and grandeur hit your senses and shake your very soul. At Gaza, for instance, there is a massive structure of rock that rises sheer for several thousand feet. Perched above it is a strange formation of red crags shaped like a gigantic cathedral, outlined against the blue sky. From the base of the walls long sweeping slopes of blue and slate-grey rubble run down to the Spiti river. These alternate with grassy precipices, steep and menacing. In the crevices up above, which look like the ramparts of a fort, lies pure white snow. Beyond these, through a gap in the fortifications, you see a snow-covered range sprinkled over with wisps of bright cloud. There lies Shangri La, the land of beauty, sunshine and perpetual youth.

If only one could have gone on looking at this dream-vision and forgotten everything else ; forgotten that in striving to reach this point we had to labour up steep hills step by step, straining and sweating in the midday heat ; forgotten the moments of anxiety and vertigo when a glance at the torrent, deep down in the valley below, induced a sense of fear and an almost irresistible impulse to jump ; forgotten the numbing fingers of cold rain tracing a pattern of pain down our backs, and the hot sun oppressing body and soul. If one could forget the hard, cruel impact of Nature on puny man and remember only her gaiety and splendour. But man is always shaken out of his moments of ecstasy by a rude shock, and then he knows that Nature is completely indifferent to his little thrills of aesthetic pleasure and his mental orgasms. The mishap that befell us on our first evening in Gaza was no more than a word of warning that what man really needs and longs for is warmth and soft comfort.

We were sitting in front of the mess-tent and chatting idly about the peace around us and the true functions of good food. The evening air was chilly, and Shrinagesh sent for his warm pullover. One of our men came up running and said that a muleteer had been drowned in the river and his horse had been washed away. We ran down towards the river, and near the muleteers' camp saw a man bringing up a bundle on his back. He unloaded it on the ground, and sat down beside it in silence. We looked at the heap of cold, inert flesh and shuddered. But the man was not dead. We straightened his limbs, turned him over on his face, administered artificial respiration, and gave him a vigorous rub all over. After half an hour of this treatment he began to show signs of life, and groaned. A few minutes later he recovered consciousness, and the first thing he did was to enquire about his horse. He was told to keep quiet and rest. We gave him a stiff dose of brandy, and wrapped him up in blankets with a hot-water bottle.

There was great commotion in the camp, and the muleteer's misadventure was related a dozen times. What had happened was this. In spite of a warning given by us, the mule-men went down to the river to try and take their animals over to the other side where the grazing was more luscious. Two men were to accompany the animals ; and as the water was too deep for fording, they mounted their horses and rode into the stream. The train of mules followed. Some of the animals swam across and reached the other side. The mounted ponies got into difficulties and one of them stumbled and fell, throwing his rider into the water. The man, in a desperate effort to save himself, clutched at the leg of the second rider, who was also dislodged from his seat. The first man succeeded in swimming across to safety, but the second, who was taken completely unawares, was washed into the middle of the stream where the current was fastest. He was carried down for a hundred yards, while he strove to extricate himself from the powerful grip of the current and shouted for help, till he hit a rock and lost consciousness. Two men who had seen him fall ran down the bank

and past him ; they caught him as he came spinning
and bobbing up and down like a piece of flotsam, and
lifted him out of the ice-cold water. The wretched
man was chilled to the bone and all but drowned. The
horse was irretrievably lost.

During the night the man developed a high temperature and we feared he had contracted pneumonia. All
we could do was to give him a few tablets of sulpha-
diazine, and tell him to stay behind and rest till he recovered. He looked very ill, but a week later, when
he rejoined us on our return journey, he had regained
his normal vigour and cheerfulness.

The 10th of July was a day of rest, and we breakfasted leisurely at nine o'clock. Shrinagesh wanted to
climb up the hill behind the village and see the asbestos
mine. He set off at ten o'clock, accompanied by
Bachittar Singh and a small army of villagers, ponies
and yaks. He expected to be back for lunch, as the mine
was reported to be only two miles away. He returned
at five o'clock after a most exhausting day. The mine
was more than six miles from the camp and involved a
strenuous march of four hours. It yielded asbestos of
good quality, but the quantity available and the situation of the mine made it economically unprofitable,
and its working had been abandoned. Shrinagesh
showed me a small leaf of asbestos which he had peeled
off the wall of the mine, and stored it away as an exhibit
forming part of his official report on Spiti.

I spent the day reading, supervising the laundering of
our linen by one of the villagers, and inspecting the
court of the Nono. The Nono is the descendant of the
titulary chiefs of Spiti. Chhewang Tobge, the present

incumbent of the post, has no real power or authority, though he enjoys certain privileges granted by the government, and commands a certain amount of respect and influence in the valley. Under the Spiti Frontier Regulation of 1878 he exercises magisterial jurisdiction, and can try all offences except murder. But the only sentence he can award is a sentence of fine. He cannot deal with civil disputes. These must be heard by the Assistant Commissioner or by the Subordinate Judge, of Kulu. The Nono is also responsible for the collection of the land revenue. In this work he is assisted by a kind of privy council, consisting of village headmen who receive the revenue from the cultivators before passing it on to him, and also advise him on a variety of matters. The headmen are appointed by the Nono in consultation with the villagers. The appointments require ratification by the Assistant Commissioner. Another privilege conferred on the Nono is the right to cultivate some government land in Dangkar village and take its produce. In return he is required to keep the Dangkar fort in repair. His personal holding, which is not inconsiderable, is situated in his home village, Kyuling.

In appearance Chhewang Tobge is a tall, handsome individual. He dresses more elegantly than the common Spiti villager, his cloak is made of finer material, his cap is trimmed with better fur and his mount is larger and statelier and more richly caparisoned than the other ponies. A Chinese carpet-piece of thick pile and bright colours, laid over his saddle, adds gaiety and distinction to his horse, and I saw that his stirrup-irons were made of carved steel and beautifully shaped. But despite these trappings of office and dignity, the Nono is what is usually called 'a dope' in more senses than one. He knows no Urdu or Hindi and could not talk to us without an interpreter. Even in his own language he is sparing of speech and we seldom saw him talking to anyone but the Tashi Nono, a kind of Pooh-Bah who acted as his deputy, clerk, interpreter and general fac-totum. The Nono had accompanied us from Losar, but had kept very much to himself. I learnt later that there were two very good reasons for his shyness. Shrinagesh and I always walked, except for brief distances when fording a stream or climbing a steep hill towards the end of the day's march. The Nono was reluctant to walk, both because he did not wish to tire himself, and because it is a gross indignity for the Nono to be seen trudging along on foot. Secondly, if he remained with us he would be condemned to long periods of abstin-ence, for we knew from his bloodshot eyes, his lisping tones and a strong alcoholic odour that he remained continuously in a state of semi-inebriation. *Arak* and *chhang* played a very important part in his life.

My chief reason for making an official inspection of his work was a complaint that he had been exceeding his powers and inflicting sentences of imprisonment in some cases. Not content with passing orders which were clearly illegal, he had carried them out in a most barbarous manner. The offenders, after conviction, were incarcerated in a horrible cell at Dangkar, which consisted of a small room built on the top of a tower of rock. Two sides of this cell were enclosed by a natural precipice several hundred feet high, and on the remain-ing two sides strong walls of stone had been built.

There was no door to this room, and a small hole looking over the precipice admitted a few rays of light and served as a ventilator. The prisoner was lowered into this dark hole through a small opening in the roof which was immediately covered with a large stone, and in this dungeon he served out his sentence, alone and unattended. His food was supplied by the complainant, who lowered it down to him once a day through the opening in the roof. The rations, it may be expected, were scanty, and not infrequently the complainant was guilty of gross negligence in discharging his obligation.

I was assured by every one I questioned that this practice had ceased, but my enquiries left me with an uneasy suspicion that a sense of loyalty to the Nono might have inspired a conspiracy of silence. I had no means of determining the truth of the matter. However, I consoled myself with the thought that nobody will suffer in silence beyond a certain limit, and had there been instances of cruel treatment the people would most certainly have spoken to us, for they knew that we represented the power of the State.

The Nono's court had no fixed location. He tried cases at the spot and dispensed a kind of rough and ready justice. Fortunately the Spitials are a peace-loving people and serious crime is a rare occurrence in the valley. During the years 1948 and 1949 the Nono had heard only eight cases of a petty nature. I turned over the pages of his register and read the entries : stealing a cloak — fine Rs.5 ; causing simple hurt — offence compounded ; stealing a sheep — fine Rs.10 ; seducing a married woman — accused acquitted. . . . There was not much to see in the court files. The Tashi Nono did all the work. He drew up the complaint at the instance of the aggrieved party ; he recorded the complainant's statement, wrote out the orders of the court, took down the depositions of the witness and the plea of the accused person, and after the conclusion of the hearing he wrote out a judgment which the Nono pronounced. The Nono himself was little more than a silent spectator of the entire proceedings. I pointed out to him that this was highly improper, and that he must learn to read and write Urdu. Two years earlier he had given an undertaking to this effect, and if he wanted his powers to continue, he must really sit up and pull himself together, etc., etc. . . .

Two years after my admonition the Nono was just as innocent of the Urdu language as before. But now there are schools at Losar and other villages ; and a good bridle-path, along the Chandra river and over the Kunzum pass, is under construction. Soon there will be bridges over all the torrents, and the bridle path will, in time, become a jeepable road. Spiti will no longer remain an isolated pocket of shy and backward humanity. But the ways of the civilized world do not promote peace and contentment, and we may find that the Spitial has lost his air of serene joy and his sense of innocent satisfaction with the niggardly portion which Nature doles out to him.

The day's rest had done me a world of good, and I felt fresh and buoyant in the morning. We breakfasted and packed up, and were ready to start at seven, but the mules were still across the river. We went down to the water's edge to watch Operation Recovery. Two mule-men were, in vain, coaxing and urging the animals to enter a swirling torrent of angry yellow water. The mules lowered their snouts to the water, sniffed at it and turned back each time they were pushed forward. The men moved the animals up and down the bank, and tried several places with the same result. At last the mules allowed themselves to be driven into the water at a place where the stream was comparatively narrow. They had to swim a distance of about twenty yards where the water was deep, and in negotiating this portion they were carried down nearly a hundred yards ; but we saw them come out of the water, one by one, and heaved several sighs of relief. At twenty minutes to nine we were on our way to the Lingti camping-ground.

We had decided to halt at Lingti in preference to Dangkar, three miles farther, because of its advantage as a more suitable starting-place for the next day's march. It was a wise decision, for Lingti is a small and restful pocket of green grass and little springs that lie concealed under bunches of moss and loose stones. A little digging with the hands and a moment's rest for the earth to settle down were enough to provide an adequate supply of clear, cool water. On three sides of this fairy nook tall mountain walls gave shelter and a picturesque setting to our camp, and on the fourth the Spiti river rumbled and groaned in the depth of the valley.

Shrinagesh and I were the first to arrive at this spot, and we dug up a new spring from below a stone, because a tell-tale pattern of hoof-marks round the old one showed that it had been fouled by mules and yaks. After a little while, Shrinagesh suggested a visit to the

Dangkar fort. I was tired, for the day's march had been hot, but I did not want to miss the opportunity of seeing Dangkar and the Nono's prison cell. So, mounted on ponies, we started on what seemed then the longest three miles I had ever travelled. After crossing the Lingti stream below our camping-ground, by a foot-bridge, we came on to the face of a barren ridge that swept round in a huge semicircular arc dominating the valley. The air imprisoned in this gigantic amphi-theatre was still, and the sun beat down in all its fierce-ness through the rarefied atmosphere of 12,000 feet above sea-level. The whole mountain was parched dry and the sun's heat came back from it in slow, tremulous waves that made the landscape quiver. Our ponies moved forward, panting and labouring up the steep and narrow path. A low cloud of dust rose up from the ground and hung in the air all along our trail. No one spoke. From time to time we raised our eyes to look at a cluster of ant-hills standing on a high mound at the end of the semicircle. This was Dangkar Fort, and after an hour it was no nearer than when we started.

Shrinagesh's horse stopped every few yards and needed a great deal of coaxing and urging before he resumed his slow march. After a time he declined to move forward. I saw Shrinagesh dismount and con-tinue his journey on foot. A few minutes later my horse began to scratch the ground with his foreleg. I pulled the reins, dug my heels into his belly and made clicking noises with my tongue. He took two steps forward and sat down. The Nono said the horse was tired and would go no farther. I stepped off his back and sat down on a stone. I told the Nono to go on, and said I would wait for the Commissioner Sahib till he returned from Dangkar, or go back to the camp. The Nono offered me his own mount, and this carried me another half-mile up to a fresh-water spring where I had several long drinks of cold sweet water. It was a heavenly spring and I drank till I could drink no more.

Dangkar is a quaint village situated at an altitude of 12,774 feet, on the summit of a steeple-shaped hill which stands out from the main ridge. The old fort is situated on the point of the steeple, while the houses go chasing each other down the steep declivity, or playing hide-and-seek among blocks and columns of hard conglomerate growing on the hill. The arable land of the village lies far away, down almost at the bottom of the valley, and the inhabitants of Dangkar must spend a great portion of their lives journeying up and down the hill. There are no trees or bushes nor any signs of verdure on the hill, which stands arid and barren like a mighty ant-hill, rising sky high from a patchwork of green fields. The fort itself is a confusion of dark passages, slippery stone staircases and filthy rooms. The Nono's agent served us with milk curds spooned out of an earthen pot, of which the outside (and, I have no doubt, the inside) had a thick coating of grime. I was unwilling to poison myself, and, for once, my *noblesse* did not oblige.

We paid a brief visit to the monastery, but the adventure proved unexciting. The prison cell, however, was every bit as impressive as the description we had heard. We could find no evidence of its having been used recently.

Dangkar was, in olden days, the capital of the Spiti

province. Its position invested it with a certain measure
of security. It is almost impossible to storm the fort
and dislodge the person in possession. Towards the
end of the 18th century Bashahri raiders captured the
fort and held out against the government for two whole
years. Spiti was annexed by the British Government
in 1849, and Major Hay, Assistant Commissioner of
Kulu, took charge of the area. He spent the winter of
1849 in the Dangkar fort, and wrote an account of his
stay there.

VI

THE PIN VALLEY

WE were back in camp before seven o'clock, and after a cup of tea I went across to the Bhavnanis' tent to have a chat. They said the muleteers had fears about the next day's march. The villagers who had come from Dangkar to help our men had been saying that the path to Kuling was not safe for laden mules. According to the official gazetteer it was 'most dangerous'. We had seen some difficult patches on the way to Lingti, and I thought that if the mule-men were afraid, the path to Kuling must be bad indeed. While we were talking, the muleteers came in a body to make a representation to Shrinagesh. He assured them that men from the Pin valley had been working on the path and the route had been much improved. He said he would walk in front and see that the mules went through safely. The mule-men looked at each other, murmured dissatisfaction and retired. I told Shrinagesh that after the mishap at Gaza it would be unwise to take unnecessary risks, and he would be well advised to abandon the idea of visiting the Pin valley. I said:

'The mule-men are obviously scared and don't want to lose any more animals or men.'

Shrinagesh thought I was showing signs of weakness and cowardice. He was short and decisive:

128

'Let those who feel doubtful stay behind. I shall go on. I shall certainly go on.'

Saying this he went away to his tent, like a tragedian making a spectacular exit. I was both piqued and amused, but said nothing, and reserved my retort for a more appropriate occasion. If the fears of the mule-men were justified, as indeed they must be, I should be able to say . . . well, I had a whole night to think out what I should say — something soft and crushing, not merely a smug 'I-told-you-so', the *mot juste* for an occasion of this kind.

But the opportunity never came.

We were ready at half-past seven in the morning, and as soon as the loading of the mules began, Shrinagesh and I went down to the suspension bridge over the Spiti. The bridge was built in 1911 against a cliff, to avoid the danger of avalanches; but it had suffered damage from falling rocks, and bore the impress of forty years' weathering. At the far end, some of the boards were missing, and there was a hole large enough for a laden mule to slip through. We had sheaves of rose bushes placed over the gap, and surfaced them with stones, grass and earth. Strangely enough, the mules were able to cross without a mishap. They went over, one by one, sniffing at the worn-out boards, while the bridge shook and trembled ominously at each step.

A hundred yards farther, the path was so narrow and lay so close to the cliffside that it was impossible for a laden mule to pass. While the baggage was being unloaded and carried across this part of the route, Ashok and I went ahead. After a short distance, it was easy going, and the path went down to the river-bed. There was no tree or bush or projecting rock to give shelter from the sun, and we sat down on the stony floor to wait for the mules. There were fifty of them and it took two hours to unload them, lead them across the narrow ledge in the cliffside and load them again. Shrinagesh was in a determined mood ; he stood supervising the manœuvre, and telling the men to hurry. At half-past ten we were on our way again, and in half an hour we arrived at the mouth of the Pin gorge. Now began the portion which the muleteers had feared and the gazetteer had described as 'most dangerous'.

The Pin river rushed out of a narrow passage, roaring and swirling between rocky walls a thousand feet high and rising smooth and vertical from the edges of the seething torrent. The noise of the water lashing itself against the walls echoed and re-echoed till it swelled to a deafening volume. We could not speak to each other, and caught in this narrow world of noise and fear and heat, went forward with slow, hesitant steps. An occasional notch in the wall, or a boulder which had escaped the fury of the torrent and come to rest against the rocky bank, provided a precarious foothold. A thin coat of slime covered the boulders, and as I jumped, crawled and slithered across fifty yards of this impossible route, I knew that the mules would never be able to come, laden or unladen.

I stopped to look back. I saw the head of a mule peeping from behind a projection in the wall. The head remained poised for a moment and then lurched forward. I could not bear to see what would happen, what must happen. I turned away and hurried on. There was no need to prove the obvious.

In another fifty yards the gorge opened out and there was a broad bank above the water's edge. In front of me the path went winding and undulating over safe and easy terrain. I sat down on a stone and began to stare at the water. Immediately below me it was swaying with a gentle movement and lapping against the rocks. Ten yards away it was a seething, curdling mass of whirlpools and vortices. The clamour of its angry roar came back to me as loud as before.

Suddenly I saw a mule come up and go past me on the way to Kuling. A few seconds later another one came up, and soon the entire caravan had passed before my astonished eyes. Shrinagesh came up, his mouth rounded into a small o. I could not hear the air he was whistling, but when he came near me he burst into song and loudly proclaimed his future plans :

> If ever I marry again
> She must be just about twenty-three,
> Fresh as a daisy and crazy for me. . . .

I jumped up and joined him. We both laughed and began to walk in step.

> With tons and tons of money,
> And never give me pain ;
> And I shall be the boss of the house,
> If ever I marry again.

The sun was right overhead and was drawing all the juice out of our bodies. We walked less than two miles in an hour and felt ourselves becoming dehydrated inside and outside. I began to talk of cider, cold and plentiful in a large cask, and a tubful of shandy with an iceberg floating in it. One could drink it, bathe in it, lie down in it.

K

'I wouldn't mind if I drowned in it.'

At half-past one we came to a huge rock standing on the edge of a sandy bank. A spring of fresh cold water issued from below the rock. I lay down on the sand and put my lips to the water.

When we resumed our journey after lunch, a party of men, led by the headman of Siling, the first village in the Pin valley, met us. They offered us their ponies. Nothing loath, we accepted them, and rode all the way home to Kuling.

I had occasion to ride the local horses several times during the trek, for though we had brought six ponies with us from Kulu (including one left with Dr. Massey at Losar) Shrinagesh and I nearly always walked, with the result that our mounts were seldom available : either they had gone on ahead or were too far behind, or were being ridden by other members of our party. Except for a few days when we were approaching and leaving Spiti, the local headman and a band of villagers always attended on Shrinagesh ; and there was no lack of riding ponies.

It is an unforgettable experience to ride one of these animals. Our own horses were small enough, but those we rode in Spiti must have been cast in a special, diminutive mould. It is impossible to get a grip on them with the knees. The stirrups are worn very high so that your knee-caps are on a level with the pommel of the saddle and there can be no question of making a direct contact with your mount. If you complain and point out that *really* the horse is too small for your size, your Spiti friends will nod agreement, put on a look of deep concern and offer to bring a bigger pony, which means a pony of the same size, made to look bigger. This is achieved in a very simple way. Over the saddle are slung a pair of large saddle-bags containing grain. These increase the effective girth of your mount. Over the saddle-bags are arranged half a dozen blankets and *namdahs*, folded into small squares ; and over everything a carpet-piece is fixed by means of ropes passing round the horse's belly. This artfully consolidated and colourful animal is then placed before you, and you are asked to climb on to its back. You will, after a brief struggle and a deal of exertion, succeed in scrambling up to the summit, and establishing yourself there ; but the moment you start moving you are filled with a sense of complete insecurity. The horse seems out of reach, and you sit on a mountain of blankets which might at any time be dislodged from their eminence. This feeling of insecurity is further intensified when the horse walks down-hill, or steps quickly down the high bank of a stream to ford it. You try to grip him with your knees, stand up in your stirrups, clutch desperately at the pommel and pray to God, all at the same time. Suddenly, there is a violent upheaval, and you are thrown back ; for the horse, having negotiated the stream, is beginning to scramble up the opposite bank.

But you soon learn that it is not horsemanship that is needed in Spiti ; what you need is an emotional, a religious, almost a fanatical belief in your mount and in his ability to carry you safely across anything and everything. The Spiti horses have a conscientious objection against the knee-grip ; you must grasp them with the inside of your calves, and give them full rein. They

will go up or down, on the level, through water, over
stones and pebbles, and through a maze of boulders.
They will pick their way along a two-inch track, lying
across the face of a steep and slippery cliff, get a firm
foothold where there is no room to place a goat's hoof,
and go on performing miracles of balance, stability and
endurance.   After the fiftieth miracle you feel you
could trust your body and your soul to the insignificant
quadruped who looks hardly strong enough to carry a
load of 20 lbs.   One of these animals transported my
186 lbs. over a distance of four miles on that hot day
in the Pin valley.   There were moments when I felt
nervous, even afraid, but my horse never faltered, never
hesitated, never slipped nor took a false step.   I could
only wonder and, in the end, feel grateful.

We camped in the sandy bed of the river, at the foot
of a cliff above which lay the Kuling village.   A crowd
of people had come down to greet us, and almost as
soon as we arrived, a party of three *buzhens* asked for
permission to entertain us.   The mules were not yet in
sight and we sat down on a flat stone to listen to the
music of the *buzhens*.   One of them played, with a bow,
on the strings of a small instrument, shaped rather like
a mandolin, and sang a long plaintive ballad in praise
of the Dalai Lama.   The other two rotated prayer-
wheels, and joined in the chant.   The performance
lacked vigour and feeling, and we listened to it out of
politeness at first, and then because there was nothing
else to do.   We were told that in the morning a whole
troupe of *buzhens* would be arriving to give a display
of dancing, singing and magic.

The *buzhens* are a peculiar feature of the Pin valley.

The lamas of Pin belong to an order which permits matrimony. The monks themselves live in the *gompa*, but their wives and children reside in the village. The sons of the lamas join the order of *buzhens*, who are a kind of peripatetic monks, though it would be more appropriate to call them strolling players. They wander about the country in small groups, earning their living by singing, dancing and acting plays. Some of them trade in a small way by bartering grain for salt, iron or honey. Unlike the monks of other orders, they do not shave their heads, and wear their hair plaited behind them or in long straight twists. Sometimes it falls about the head in an untidy mop giving the *buzhen* a wild look. Eighty years ago a grand lama from Tibet paid a visit to Lahoul and Spiti and expressed strong disapproval of their unclerical appearance. He ordered the *buzhens* to cut off their hair, but his admonition had little effect, and the strolling monks retain their hirsute appearance to the present day.

It is said that the *buzhen* order was founded by one Thang-teong-Gyalpo (king of the desert), because the people of Tibet were corrupted by Langdarma, the famous king of Lhasa. Langdarma won them over from Buddhism and made them follow a religion of his own. He succeeded so well that in the course of fifty years the old faith was quite forgotten and the sacred *mantra: Om manipadme hum* was no longer heard in the country. Appalled by this state of affairs Chan-re-zig, the deity worshipped at Triloknath in the Lahoul valley, came down to earth to redeem the people. His incarnation was born in the king's house as Thang-teong-Gyalpo. The child grew up to be a saint and a reformer, but soon found that the people had progressed so far on the path of wickedness that they could not be drawn back by scriptures and dull sermons. He therefore adopted the dress and manners of a wandering minstrel, collected a band of followers, similarly attired, and began to wander from village to village, offering to amuse and entertain the people by performing miracle plays, on condition the audience repeated the chorus *Om manipadme hum* whenever it occurred in the chants or recitations. In this way the people once again learnt the power of the sacred *mantra*, their mouths became purified, and the religion of Buddha was revived.

The miracle play acted by the *buzhens* has a constant theme. Except for small variations, peculiar to each company of players, the story is this. One day a certain anchorite who had lived alone for twelve years in an inaccessible forest washed his robe in a pool in the hollow of a rock. A doe drank the water in the pool and conceived therefrom. In due course it gave birth to a baby girl at the door of the anchorite's cell. The girl was adopted by the hermit and grew up to be a beautiful woman. She was named Sun-face and married a king. The other queens in the palace became jealous of her and conspired against her. They accused her of being a witch and of eating human flesh. They murdered her child, and, showing its dead body to the king, said that Sun-face had killed it to feast on its flesh. The king believed this story and turned Sun-face out of his palace. She wandered about the forests for several years, till the king discovered the plot, put the conspirators to death and recalled her.

Our minstrels continued to entertain us till the tents were put up and the tea was laid. We had had a tiring day, and after tea I lay down in my tent and read Bridges' *Testament of Beauty*.

How small a part
of Universal Mind can conscient Reason claim!
'Tis to the unconscious mind as the habitable crust
is to the mass of the earth ; this crust whereon we dwell,
whereon our loves and shames are begotten and buried,
our first slime and ancestral dust : 'Tis, to compare,
thinner than o'er a luscious peach the velvet skin
that we rip off to engorge the rich succulent pulp :
Wer but our planet's sphere so peel'd, flay'd of the rind
that wraps its lava and rock, the solar satellite
would keep its motions in God's orrery undisturb'd.

That set me thinking about Buddha's teaching that ignorance is the root cause of all pain and suffering, and it is only when the conscient mind reaches the state of full knowledge that release is obtained. The Christian must expiate the Original Sin of his primeval ancestor, by worshipping God, and believing in Christ as the Saviour of mankind. He must follow the path of good deeds and shun wickedness. If he does this he will receive his reward on the Day of Judgment and his soul will enter the Kingdom of Heaven. The wicked man will be cast out and thrown into the outer darkness. His soul will for ever remain in a region where there is perpetual burning and wailing and gnashing of teeth. For the Moslem, belief in the Prophet Mohammad ensures salvation — the true believer goes to a heaven full of luscious houris, disporting themselves in a garden where there is abundance of green grass and water and trees laden with succulent fruits. The unbeliever, the *kafir*, goes and burns in hell. The reader of the Bhag-wadgita is told to work out his salvation by good deeds and exclusive devotion. 'Do not run away from the world,' says the gospel of Lord Krishna, 'face the difficulties in your way, and do your duty, even if it involves the killing of your dear ones. Do not look for merit when you do a good deed ; if your motives are good, your deed is righteous. Therefore do your duty, pleasant or unpleasant, unflinchingly. Be my *bhakt*, my devoted worshipper. I shall give you salvation.'

Krishna, Mohammad and Christ all taught that there is one God, Eternal and Everlasting ; the soul of man is immortal and imperishable and salvation comes after death. The reward which a righteous man hopes for is deferred, and the punishment of the evildoer is delayed. Whether you are reborn on earth to work out your *karma* in the next incarnation, or your disembodied spirit goes to heaven or hell, your reward will come in the after-life. The secret teaching of the Upanishads was only a little less despairing. You could release your soul from bondage by obtaining knowledge of Brahma, the Creator, and realizing that your own self or ego was identical with him. Then your soul would leave your body and go to join Brahma and become one with him.

The follower of Buddha does not crave for a future life of ease and comfort, nor does he feel complacent about his immunity from punishment in this life. His pain and suffering and his salvation are here, in this life. There is no God who sits in judgment upon him, nor is his soul imperishable. He will not burn for ever in the fires of hell, he does not look forward to wallowing in a heaven of sensuous delight till the end of eternity.

He must seek knowledge and destroy ignorance, and thus free himself from pain and suffering and evil.

The teaching of Buddha was direct and simple. Shorn of its deistic and metaphysical growth which later surrounded and all but changed its aspect, though not its essential character, it may be stated thus :

There are three fundamental principles or axioms—

(i) There is nothing permanent or everlasting ; everything is in a state of continuous change. As soon as a thing begins to be, it also begins to end. This is the Great Law of Impermanence or the Doctrine of *Aniccam*.

(ii) The origin of sorrow is the origin of individuality. When an individual begins to be a separate entity, the outside world makes an impact on his senses, and his sensations are stirred up. Desires arise, and to satisfy them he makes an effort to remain separate from the rest of existence. But since his desires cannot be completely satisfied, pain and sorrow result. This is the principle of Sorrow being inherent in Individuality, or the Doctrine of *Dukham*.

(iii) The notion that there is an ego or a separate individuality is a delusion. It is an error to think that man is anything more than a mere link in a long chain of causation. He cannot think of himself as an independent being apart from everything else. This is the principle of Non-reality or Impermanence of the separate ego, or the Doctrine of *Anattam*.

This is how Buddha enunciated the threefold doctrine to his disciples :

Whether Buddhas arise, O priests, or whether Buddhas do not arise, it remains a fact and the fixed and necessary constitution of being, that all its constituents are transitory.

This fact a Buddha discovers and masters, and when he has discovered and mastered it, he announces, teaches, publishes, proclaims, discloses, minutely explains, and makes it clear, that all the constituents of being are transitory.

Whether Buddhas arise, O priests, or whether Buddhas do not arise, it remains a fact and the fixed and necessary constitution of being, that all its constituents are misery. This fact a Buddha discovers and masters, and when he has discovered and mastered it, he announces, teaches, publishes, proclaims, discloses, minutely explains, and makes it clear, that all the constituents of being are misery.

Whether Buddhas arise, O priests, or whether Buddhas do not arise, it remains a fact and the fixed and necessary constitution of being, that all its elements are lacking in an Ego. This fact a Buddha discovers and masters, and when he has discovered and mastered it, he announces, teaches, publishes, proclaims, discloses, minutely explains, and makes it clear, that all the elements of being are lacking in an Ego.

Once these principles are accepted and understood, the meaning of the Doctrine of Causation as expressed in the Four Noble Truths becomes plain :

(i) Suffering is omnipresent and inevitable.

(ii) Its cause is the folly of believing in a separate ego and selfish desire.

(iii) The remedy lies in the elimination of ego and selfish desires.

(iv) The remedy is procured by following the Noble Eightfold Path. The Eightfold Path is the path of (1) Right views, (2) Right thought, (3) Right speech, (4) Right conduct, (5) Right means of livelihood, (6) Right effort, (7) Right mind control, and (8) Right meditation.

Buddha taught that the way to seek salvation is not

through a life of pleasure or a life of extreme austerity, but by following the Middle Way — the Noble Eightfold path. His own life of pleasure in his father's palace, where he was surrounded by every kind of luxury, did not bring him peace and freedom from pain ; nor did he gain anything by fasting and practising austerities. In his first sermon to the five recluses who attended on him shortly after his enlightenment, he explained the fundamental principles of Buddhism and indicated the means of escaping from sorrow and pain :

These two extremes, O monks, are not to be practised by one who has gone forth from the world. What are the two ? That conjoined with the passions, low, vulgar, common, ignoble, and useless, and that conjoined with self-torture, painful, ignoble, and useless. Avoiding these two extremes the Tathagata has gained the knowledge of the Middle Way, which gives sight and knowledge, and tends to calm, to insight, enlightenment, Nirvana.

What, O monks, is the Middle Way, which gives sight . . . ? It is the noble Eightfold Path, namely, right views, right intention, right speech, right action, right livelihood, right effort, right mindfulness, right concentration. This, O monks, is the Middle Way. . . .

(1) Now this, O monks, is the noble truth of pain : birth is painful, old age is painful, sickness is painful death is painful, sorrow, lamentation, dejection, and despair are painful. Contact with unpleasant things is painful, not getting what one wishes is painful. In short the five *khandas* of grasping are painful.

(2) Now this, O monks, is the noble truth of the cause of pain : that craving, which leads to rebirth, combined with pleasure and lust, finding pleasure here and there, namely the craving for passion, the craving for existence, the craving for non-existence.

(3) Now this, O monks, is the noble truth of the cessation of pain : the cessation without a remainder of that craving, abandonment, forsaking, release, non-attachment.

(4) Now this, O monks, is the noble truth of the way that leads to the cessation of pain : this is the noble Eightfold Path, namely, right views, right intention, right speech, right action, right livelihood, right effort, right mindfulness, right concentration.

As long as in these noble truths my threefold knowledge and insight, duly with its twelve divisions, was not well purified, even so long, O monks, in the world with its gods, Mara, .Brahma, with ascetics, brahmins, gods and men, I had not attained the highest complete enlightenment. Thus I knew.

But when in these noble truths my threefold knowledge and insight duly with its twelve divisions was well purified, then, O monks, in the world . . . I had attained the highest complete enlightenment. Thus I knew. Knowledge arose in me, insight arose that the release of my mind is unshakeable ; this is my last existence ; now there is no rebirth.

In travelling along the Noble Eightfold Path, the seeker after knowledge and enlightenment must, one by one, break and cast off the ten fetters or *samyajans* which hold him in bondage and keep him prisoner in the region of ignorance. These fetters are :

(i) The delusion of self or the belief in a separate individuality ;

(ii) Doubt or lack of faith ;

(iii) Belief in the efficacy of good works and religious ceremonies ;

(iv) Bodily passions ;

(v) Ill-will or malice ;

(vi) Desire for individual life in the world of form or attachment to tangible objects ;

(vii) Desire for individual life in the formless world or a craving for a life in heaven ;

(viii) Spiritual pride ;

(ix) Self-righteousness ; and

(x) Ignorance.

When the last fetter is broken and thrown away, the traveller reaches his goal, and attains the state of Nirvana, the 'going out', when the threefold fire of lust, ill-will and delusion is extinguished ; when the desire to grasp has gone and there is peace and purity and complete emancipation. This is the state of which Buddha speaks in the following terms :

There is the stage, where there is neither earth nor water, nor fire, nor wind, nor the stage of the infinity of space, nor the stage of nothingness, nor the stage of neither consciousness nor non-consciousness, neither this world, nor the other world, nor sun and moon. There, monks, I say there is neither coming nor going, nor staying nor passing away, nor arising ; without support or going on or basis is it. This is the end of pain.

This is a convenient place to say something about *bhavachakra* — the Wheel of Becoming, or as it is sometimes erroneously called, the Wheel of Life. The twelve divisions of knowledge and insight mentioned by Buddha in his first sermon are pictorially represented by a six-spoked wheel. The six destinies, hell, animals, ghosts, gods, the rebel gods and human beings are shown in the spaces between the spokes. In the centre are represented passion in the form of a dove, hatred in the form of a snake and stupidity in the form of a pig.

The causal formula is stated in a series of twelve pictures painted along the circumference. These represent the twelve *nidhanas* — occasions or causes — or the twelve divisions of knowledge and insight :

(i) *Avidya* (Pali, *avjjia*), ignorance. This is represented by a blind man feeling his way with a stick.

(ii) *Samskarah* (Pali, *Sankhara*), literally confections or aggregates, in other words all the immaterial qualities of thought, word and deed which make up an individual. These are shown by the picture of a potter working his wheel, with pots lying around him.

(iii) *Vijnana* (Pali, *Vinnana*), consciousness represented by a monkey climbing a tree with flowers.

(iv) *Namarupa*, lit. name and form, represented by a boat with four passengers, crossing a river. The boat is the body of man and the passengers in it are the four *skandas*, or immaterial group-feelings, perceptions, *samskarahas* (*vide* (ii) above) and consciousness who is steering the boat.

(v) *Shadavatanani* (Pali, *salavatana*), the six provinces or territories — the five senses and the mind — shown by an empty house with six windows.

(vi) *Sparsha* (Pali, *phasso*), contact, represented by a man and woman embracing.

(vii) *Vedana*, feeling or sensation, represented by a man with an arrow entering his eye.

(viii) *Trishna* (Pali, *tanha*), thirst, craving, shown by the picture of a man drinking from a pitcher held by a woman.

(ix) *Upadana*, grasping, meaning man's attachment to worldly things, represented by a man gathering fruit from a tree, or picking flowers.

(x) *Bhava*, becoming, or the tendency to come into being, pictured as a pregnant woman.

(xi) *Jati*, birth, represented by a woman in labour.

(xii) *Jaramarana*, old age, decay and death, represented by a corpse being carried to the cemetery.

The image of Impermanence is shown swallowing the entire wheel ; and in the top right-hand corner Buddha stands, pointing at the Wheel of Becoming.

Thankas depicting the Wheel of Becoming are to be found in all Buddhist monasteries, and in the course of our trek in Spiti and Lahoul we saw several fine examples. Sometimes a simplified wheel is inscribed on a stone placed on the *mane* walls. Occasionally the wheel is five-spoked, but the six-spoked form is the one most commonly met with.

The *bhavachakra* attempts to show in pictorial form the entire chain of events as they take place in life — how one thing leads to another, and everything is inevitably tied up with the chain of causation. Ignorance (i) is the cause of individuality or individual existence (ii), made up of thoughts, words and deeds peculiar to the individual, and this gives birth to consciousness (iii) which is the first and rudimentary consciousness of being. The man is not yet a reasoning creature, possessing complete awareness of himself and his surroundings. In the next stage he possesses a name and a form (iv) ; he is conscious of himself, as he proceeds along his career on earth. He now acquires the six senses (v), and when impulses from the outside world make their impact (vi) upon his being, he feels the impact (vii), and his senses react to it. This inevitably gives rise to a craving inside him, the —

Trishna, that thirst which makes the living drink
Deeper and deeper of the false salt waves,
Whereon they float, pleasures, ambitions, wealth,
Praise, fame or domination, conquest, love ;
Rich meats and robes, and fair abodes and pride
Of ancient lines, and lust of days, and strife
To live, and sins that flow from strife, some sweet,
Some bitter. Thus Life's thirst quenches itself
With draughts which double thirst.[1]

And so comes attachment to worldly things (ix), and
the desire to grasp. This desire brings about the tendency to be, to come into existence (x). Because men
have an overpowering desire to hold on to worldly
goods, and enjoy the pleasures of the senses, they acquire
the tendency to be, to exist and thus they are born (xi)
and being born, they suffer sorrow, pain, old age (xii) ;
they decay and die. It is only when you destroy ignorance that you put an end to all pain and suffering :

In one who abides surveying the enjoyment in things
that make for grasping, craving increases. Grasping is
caused by craving, coming into existence by grasping,
birth by coming into existence, and old age and death by
birth. . . . Just as if a great mass of fire were burning
of ten, twenty, thirty, or forty loads of faggots, and a
man from time to time were to throw on it dry grasses,
dry cow-dung, and dry faggots ; even so a great mass of
fire with that feeling and that fuel would burn for a long
time. . . .

In one who abides surveying the misery in things that
make for grasping, craving ceases. With the ceasing of
craving grasping ceases, with the ceasing of grasping coming
into existence ceases, with the ceasing of coming into
existence, birth ceases, and with the ceasing of birth old

[1] Sir Edwin Arnold, *The Light of Asia*.

age and death cease. Grief, lamentation, pain, dejection, and despair cease. Even so is the cessation of all this mass of pain.

After breakfast we went out for a short walk up the Pin valley and paid a visit to Kuling. The village and the fields surrounding it are situated on a gently sloping plain which, at the time, was covered with rich green barley and masses of wild flowers. In the centre of the village there stands a *peepul* tree (*Ficus religiosa*), the only one I saw in Spiti ; and near the village is a shady grove of willows. There are not more than ten or fifteen houses in the village, and they are arranged neatly along the sides of a broad lane. A narrow channel, carrying the waters of a snow stream to the barley fields below the village, flows through what in England would be called the 'High Street'. At a short distance from the village lies the monastery where a gorgeous life-size statue of Buddha is enshrined. The figure was carved out of wood and encased in silver by the silversmiths of Rampur Bashahr, and is exquisitely fashioned.

Just beyond the village where the river-bed narrows by the jutting out of a promontory, a cable had been stretched across the stream to serve as a primitive form of bridge. We sat down on the high bank and watched the people being strung to the cable and pulled across. It was scarcely an agreeable mode of travelling across the water ; and at the far end the downward swing of the cable as it oscillated almost completely immersed the passenger several times in the stream, but it delivered him on the other side in one piece, and there was no other means by which this could be accomplished, for in

L

the summer months the Pin is quite unfordable. Shrinagesh thought the experience would be worth the discomfort of being tied to an iron cable, but while he was
making up his mind to offer his body for the exercise,
the tug-rope broke. The passenger on the cable was
safely landed, but further crossings had to be suspended.
It was nearly one o'clock, and we decided to return to
the camp for lunch.

The performance of the *buzhens* had been announced
for four o'clock in the afternoon, but soon after one a
crowd began to collect near our camp. Men, women
and children from the neighbouring villages had turned
out in large numbers to see the show. By half-past
two the space intended for the spectators was full, and
a murmur of impatient voices rose above the noise of
the torrent. At three o'clock we were told that the
performers could wait no longer, as many of the spectators had to travel long distances to reach home ; and
the entertainment must be concluded early, to give
them a timely start. We came out of our tents and
took seats in the front row on our camp-chairs.

At the back of the empty space forming the stage a
row of low tables had been arranged. On these tables
rested brass images, incense burners, *dorje* bells and other
articles of ritual. On the ground, below the tables, lay
several conch shells, gaily coloured ribbons, a number
of swords, a stick resembling a magician's wand, an
empty bottle, a brass bowl containing barley grains and
things which must have come out of a magician's bag.
Above the table was stretched a silk banner, painted in
bright colours with dragons and figures illustrative of
local legends. This served both as stage scenery and a

screen behind which the actors retired to change their costumes and make ready for the next entry.

As soon as we were seated, the leaders of the troupe came out from behind the banner, and stood in a row in front of it. Each of them was dressed in a long cloak and a head-dress consisting of a mass of streamers of multicoloured silk. One of the men blew on a conch shell while the others looked up at the sky. I thought this was an invocation to the gods, but we learnt later that it was a relic of the olden days when conch shells were blown to collect the spectators. Grains of barley were next scattered in the air, and sprinkled on the stage as an offering to the patron saint of the troupe. Finally the performance began.

The first item was a group dance, slow at first, but gradually changing to a quick and fast-moving measure. Soon the dancers were whirling and gyrating at great speed, their feet keeping time to the rapid tattoo of the drum. The two leaders were skilful dancers and their footwork was clever and fascinating to watch. The second item was a comic turn by a man dressed up as a bear-hunter. He wore a large bearskin coat, and carried a bow and arrow. After going through a series of dumb movements he began to speak like a comedian addressing his audience. He warmed to his theme, and accompanied his words with strange antics, making the men roar with laughter. The women laughed, too, but they turned away their faces and covered their mouths with their hands in coy embarrassment. We could not get anybody to translate the bear-hunter's jokes, but we laughed with the others at his obscene gestures.

The sword-dance followed. Five men appeared on

the stage, wearing voluminous skirts of brightly coloured wool, and head-dresses similar to the ones worn by the previous dancers. The torsos were bare except for a scarf. This was a brightly coloured piece of silk, measuring twelve inches by twelve, folded along one diagonal. Two ends of this triangle were pinned into the flesh behind the shoulders, so that the scarf covered the upper part of the back, and fluttered with the movement of the dance. Each dancer had a steel pin, the size of a knitting needle, with a trident at one end. This was pierced through the right cheek, so that the fork rested against the outside of the cheek ; and the point passed through the mouth, and extended for three inches beyond the left corner of the lips. The needle was pulled out and pierced through the cheek, several times in our presence, and only once did a drop of blood appear at the place where the needle had gone in.

The dancers armed themselves with a sword, held in each hand, and commenced their act. They moved slowly, sometimes in Indian file, one behind the other ; and sometimes in a straight row, advancing and retreating together. All the while they flourished their swords. The pace increased ; and suddenly one of the dancers detached himself from the others, and turning his hands inwards, applied the points of his swords to the two sides of his abdomen, just above the groin, and held them in place by pressing on the hilts with his hands, stretched out at arm's length. He ran across the stage, and with one quick movement lowered his hands, then, placing the hilts on the ground, jerked up his feet, so that for a moment his entire weight rested on the points of the two swords. The body swung back ; and the dancer, once more on his feet, went gyrating back to the end of the stage. The other players followed suit one by one, and in twos and threes. Finally all five charged forward in a row and executed the jump. The sword points were next applied to the armpits and similar movements and jumps carried out. One of the dancers pushed the point of his right sword against the inside of his left cheek, and, holding his left hand behind him, jumped sideways ; and for a brief moment remained balanced with all his weight on the point of the sword. It was really quite frightening, and when he offered to perform the feat with the points of his swords resting on his eyeballs, we called a halt to the proceedings.

The sword ends were quite sharp ; and the pressure to which the flesh was subjected, though momentary, must have been considerable ; but no injury was sustained by the dancers, beyond a faint bruising of the skin. The players, however, resented interruption ; and when Bhavnani stopped the dance to ask for a repetition of a particular movement, so that he could film it, there were murmurs of protest. When the dance was resumed it had not the same *élan*, and the dancers seemed to have lost much of their spontaneous joy ; for a few moments they revolved and hopped around like automata ; but the fervour of the miracle play once again took possession of them, and they finished in a whirlwind of emotional climax.

After a brief rest the company prepared for the last and the most important item of the day's entertainment ; and in the time-honoured tradition of true showmanship the audience were given an opportunity of showing their appreciation in the usual way. Two men went round the circle of spectators, distributing grains of barley, blessed by the *buzhens*, and receiving donations of varying amounts. Most people paid two annas ; though some gave more, and a few less. We gave a rupee each, and the players seemed well pleased with the takings. But the troupe had expectations of better things to come, and after the performance they made a formal application to Shrinagesh for a free grant of government land. Shrinagesh promised to consider the matter sympathetically and recommend the case to the government.

The money was collected, counted and put away, and the preliminary ceremonial for the stone-breaking miracle began. A part of the stage was swept clean and sprinkled with barley grains. A block of limestone thirty inches long, twelve inches broad and ten inches thick was carried forward and placed over the grains. The stone was heavy and two men carried it with difficulty. The actors all stood round the stone, chanting prayers and casting a magic spell on it. The invocation was long and elaborate, gradually increasing in pace and volume.

After a few minutes the voices trailed off, and the ring of men opened out. A small scraggy man stepped forward and lay down on the ground. He was naked above the waist ; and with every exhalation of his breath his ribs stood out prominently and outlined the frail structure of his body. A blanket was folded and laid over these fragile ribs, and over the blanket the block of stone was placed. One of the actors now brought a globular stone, the size of a football, and hit

the block with it gently, as if trying to see what result
the impact would produce. As nothing happened, he
raised the round missile, and threw it on the block of
stone. The limestone block broke in the middle, and
its two halves fell on either side of the prostrate man,
who at once jumped up, and after showing that no
harm had come to him, made a frisky exit. At the
same time the spectators rushed in, and began to pick up
the barley grains from the place where the stone had lain
during the preliminary invocation. These grains, when
added to *arak* or *chhang*, enhance the virtue and potency of
the beverage and invest it with many beneficial qualities.

The play was over and the actors were packing up
their trappings and stage properties. The spectators
were leaving for home in twos and threes, and talking
to each other in loud, unrestrained tones. To them the
miracles performed by the *buzhens* were a familiar
demonstration of spiritual power ; they did not arouse
surprise or disbelief. If you went on saying : 'Hail,
thou possessor of the jewel lotus. Amen', the ignorance
within you was destroyed, and the wisdom of Buddha
gave you peace and protection, and strength to deal
with the problems of life. The sacred intonation *om*
was a manifestation of primordial sound-waves which
filled the universe at the beginning of time, when there
was no life on earth, and the primeval cloud of elemental
dust swept through space in a chaos of swirling vortices.
O . . . o . . . m . . . m vibrating and resounding
through the world, shook the elements into order ; and
when man was born the sound was put into his mouth
for his salvation, and for the glory of Buddha — the
Enlightened One.

The actors and spectators were gone, and the river-bed in front of me was empty. A round scar in the sand showed where the miracle-makers had leapt and gambolled. In the greying light of late afternoon the wind was scattering handfuls of dust all around me, softening the impressions in the sand, dulling the black outlines of the mountains towering on all sides. There was a strange and comforting permanence about the roar of the Pin torrent ; and all other noises seemed to merge and lose themselves in the sempiternal drone which swelled and filled the valley. The river, the stones and the mountains were chanting the chorus of *Om manipadme hum.* The peace which is born of understanding and true knowledge enveloped every-thing in its warm protective embrace. Nothing seemed to matter any more, and in the heart of the Himalayas all doubts and conflicts were resolved. The fetters which bind and constrict the understanding were cast off. There was peace immanent, and awareness of joy — the joy that is the end of all craving and sorrow, the joy that is the fulfilment of life, here and now.

A cold breath of air sent a shiver down my back, and as I rose from my chair and tried to shake the numbness from my legs, I heard myself repeating the sacred *mantra* over and over again : *Om manipadme hum, om manipadme: om*, . . .

The evening light was gone ; I walked towards my tent with uncertain steps, and felt the blood once again circulating through my legs.

The night around me was black, and Bachittar Singh's radio was playing the latest film songs from Delhi.

VII

SPITI (2)

KULING marked the end of our outward journey, and in the morning we retraced our steps back to the Spiti valley. Shrinagesh and I were able to make an early start ; and by keeping up a brisk pace we reached the mouth of the Pin valley at nine o'clock. Here we met a party of geologists who were surveying the country on behalf of the Government of India. Their leader complained bitterly about the local people. He had never witnessed so much indifference, such complete lack of concern with official activities. After all, they were not tourists bent on pleasure ; they were government officers doing official work and suffering hardships for the sake of their country. Pushing himself nearly to the verge of tears, he moaned :

'They told us pack-animals could not go into Pin ; so we had to pay off our mule-men, and engage twice the number of porters and pay them Rs.5 per stage. What an unfriendly country this is. Now, when we went to . . .'

'Have you seen anything of interest, anything of commercial or industrial importance ?' we asked him.

'Oh, yes. We have collected some interesting marine fossils. The Himalayas are comparatively young mountains. The Spiti range belongs to what we call the Mesozoic period. You find samples of trias and cretaceous rocks ; for instance . . .'

I turned to the second-in-command : 'Did you find any gold, silver or other minerals, such as iron, lead, antimony ?'

He shook his head and assumed an air of defeat. There was not much mineral wealth in the region, he said. No gold or silver, very little iron, only about three per cent in the rocks at some places, hardly any magnesium. Asbestos, yes, if the problem of transport could be solved. But there was lots of limestone for cement, and clays which could be used for pottery.

This was disappointing, for I had heard tales of riches to be found in the valley. There was, for instance, the story of the army engineer who was prospecting for silver. He wandered high and low but all in vain. One day when he was traversing the rocky face of a cliff he slipped and fell. In his precipitate descent he clutched at anything that touched his hand. Half-way down he grasped a small bush ; the roots came out, and he had a momentary glimpse of silver nuggets lying in the hollow. But he fled past the vision, and came to rest only at the bottom of the cliff, where he lay senseless till he was discovered and carried to his camp several hours later. His subsequent efforts to re-discover the silver vein proved fruitless ; but he recorded his experience in the official minutes, and swore ever afterwards that a whole mountain of pure silver existed in the valley.

Our geologists had no such conviction, but they had not been so venturesome as the army engineer, and had confined their activities to the main highways and beaten tracks. It was dangerous to deviate from the narrow paths, they said.

We left the geologists to their own devices, and continued on our way. At the corner which brought the Spiti into view, we stopped to look for fossils. After digging with a crowbar for ten minutes, Shrinagesh picked up a tiny pebble with crinkly markings. It was a mean and shameful fossil, but it contained unmistakable evidence of a watery past.

Crossing the bridge near Lingti was easier, and took much less time than on the outward journey; and we were on the move again at half-past eleven. At Lidung, a small hamlet of a dozen houses, we sat under a willow tree and ate our midday meal. An hour later we were in Lara, where we camped for the night. It was a boiling hot afternoon, the temperature inside my tent was 88° F.; and when I opened my tin of *besin* (gram flour) fudge, I saw that the neatly cut lozenges had all melted and coalesced into an unshapely mass. But the taste was unimpaired; and the fudge was as satisfying and sustaining as ever.

I had asked the Tashi Nono, some days previously, to let us taste the local food when convenient, and at Lara we were served with a complete meal cooked in the local manner and with local ingredients. We were also presented with a bottle of *chhang* and a bottle of *arak*. The *chhang* had a sickly taste and a strong, unpleasant odour, not unlike the odour of bad beer. The *arak* or whisky tasted like a mixture of bad gin and methylated spirit. One sip from each bottle was as much as I could stomach; and the liquor was given away to the servants. They assured us the next day that the stuff was better than anything they had drunk before. The food was undoubtedly palatable. Thin *chappatis* of barley flour served crisp and hot and copiously buttered, *dal* of dried peas and spinach made excellent fare. To these we added a tin of Yellow Cling peaches from our own store, and said a polite 'no' to the large bowl of steaming hotch-potch which Chanchlu had sent up once again.

The next morning I borrowed Shrinagesh's hand-mirror, and examined my face very carefully. My beard was fast approaching the grizzly stage. The hair was black at the sides, and an ugly grey on the chin and upper lip. It covered the lower part of my face like a thick coating of oil colours, laid on by an incompetent hand. The brush had strayed far too near the eyes and had slipped sideways when approaching the nose.

Above the eyebrows rose a vast expanse of bareness, extending as far as the eye could see. Up to the line marking the position of the hat-rim the forehead was red and purple, and beyond it a pale pink. The nose was blotchy and multicoloured. The picture, framed in the mirror, demanded a positive reaction, and everybody said I looked like Zaffar Ullah, Pakistan's Foreign Minister ; at any rate my beard was every bit as picturesque and fearsome as his. I looked at Shrinagesh and Mrs. Bhavnani. No, none of us has the gift of seeing 'oursels as others see us'.

I took out my jar of sun-proof cream and rubbed a little of the paste on my nose and forehead.

The day was cloudy and pleasantly cool. At Gaza we enquired about the muleteer who had fallen in the river. He came up smiling, and said he had completely

recovered from the ill-effects of his accident. Dr.
Massey, however, was nowhere to be seen. Gaza was
to be his headquarters for the next three months;
and we had expected to find him well-established and
dispensing his healing balms. We were told he had
gone.

'Gone? Where?' Shrinagesh asked.

'To Kyelang.'

But how had he gone to Kyelang and when and why?
One couldn't just get up one morning and say: 'I am
going to Kyelang,' and drive there in a car. Kyelang
was nearly a hundred miles away, across two high
passes and several treacherous streams. The paths were
difficult and in places dangerous. Of course people had
travelled to Kyelang by this route before, and we were
going there ourselves, but it was necessary to have tents
and provisions and pack-animals. Our informer did
not enlighten us further. It was all very strange and
mysterious, but the next day we heard the story from
the local *patwari*.[1]

When we left Losar on the 8th of July Dr. Massey
had stayed behind to recuperate his powers and arrange
his affairs. We expected to be back at Losar after ten
days, and took with us only a part of our provisions,
leaving the rest at Losar to be picked up on the return
journey. We also left behind a few mules and two
tents. Dr. Massey spent the 8th of July in rest and
contemplation. On the morning of the 9th he announced his intention of going to Kyelang. He helped
himself to a liberal share of our provisions, collected
his medicine crates, packed up a tent and, taking four

[1] Petty revenue official.

mules and a riding pony, set off on the way to Kyelang. He chose the longer and more difficult route over the Baralacha, because he felt that nothing could be worse than his experience on the Hamta pass. It was not till several weeks later that we heard the concluding portion of this unhappy episode, and were able to add an appropriate epilogue.

Dr. Massey succeeded in reaching Kyelang, and from there he travelled to Kulu; but the rigours he experienced on the way changed him to such an extent that he wrote a long letter to the head of the medical department, making a violent attack on Shrinagesh and Bachittar Singh. He said the Commissioner and the Sub-Divisional Officer had practised deceit upon him, and had prevailed upon him to accompany the trekking party, so that a doctor should be handy, in case anyone had an accident or fell ill. He had been starved on the way, and made to walk through rain and over snow-bound passes for long periods; he had suffered bodily torture and mental anguish; he had borne everything in the spirit of Christ; but when the Commissioner rifled the medicine chests, containing government property, committed to his charge, and abandoned him in the wilderness of barren rocks and mountains, without making any provision for his comfort or safety, he could endure it no longer, and had returned to Kulu after suffering unimaginable hardships and privations. At the moment of writing he was a man broken in body and in soul, and humbly awaited orders for his future guidance.

While Dr. Massey was putting his pieces together, the Director of Health Services launched an enquiry into these allegations. The laconic wireless messages we received and sent out were not sufficiently informative, and it took some time for the truth to be known, but in the end Dr. Massey was dismissed from service on charges of defection and gross dereliction of duty. Bachittar Singh wanted to take action under the Penal Code, on a charge of theft and criminal misappropriation of a tent and sundry provisions, but he was persuaded to take a less vindictive view of the matter. I felt very sorry for the wretched doctor, and wished that he might have been awarded a less drastic punishment ; but nothing short of removal from service could emphasize the necessity of maintaining discipline.

We were to camp at Sumning across the river. A mile beyond Gaza a strange wooden contraption, bearing a faint resemblance to the ruins of a cantilever bridge, stretched across a fifty-foot chasm between two cliffs standing on opposite sides of the river. The supporting beams were made of spruce trunks which were so bent and twisted that they no longer looked like spruce trunks or supporting beams. The cross planks were nailed to the beams at varying distances, leaving big gaps in between, like flimsy sleepers fixed to a crazy railway track. The width of these planks (with one or two exceptions) was not more than six inches, and the nails which secured them had become rusty and loose. As we walked across this bridge one by one, stepping from plank to plank, the whole bridge shook and trembled, and the rickety planks rattled. Fifty feet below the wide gaps between the planks the Spiti torrent raged and foamed through the narrow bottle-neck formed by the cliffs. There was no side-railing to lend physical or moral support during the brief but perilous journey over this bridge.

Our mules were unloaded and the baggage was carried across the bridge ; the unladen mules crossed the river at a ford a hundred yards above the bridge. One of the porters slipped and fell. For a long, horrible moment he lay across a narrow plank with his legs dangling through the opening on one side, and his hands struggling to reach across the gap on the other. Shrinagesh and I stood watching his impotent movements.

'Shouldn't he drop his load ? ·Better to lose a bedding-roll than a man *and* a bedding-roll.'

'But can he ? It is tied to him, and in that position . . .'

'Good God ! Look.'

We saw the man scramble to his feet. Steady now ; slowly, old man, there's no hurry. Man and bedding hopped from plank to plank, and were safely across.

The Bhavnanis had lagged behind, and were nowhere in sight. There were several men with them, including the resourceful Tashi Nono, and they did not need our assistance. Shrinagesh and I went on to Rangrik, and paid a brief visit to the local primary school, where thirty-two young pupils and one master were engaged in a gallant but not very successful struggle with the Urdu and Hindi alphabets. The school-room was clean and tidy ; and in anticipation of our visit the students had put on clean clothes. Only two of the boys could read. Of the others, some looked at the book and recited the words from memory, some merely closed up the slits of their eyes and smiled. Shrinagesh complimented the teacher and told him to carry on the good work.

As we walked away from the school I asked Shrinagesh :

'Why didn't we ask the teacher to read something ?'

'Yes, why didn't we ?'

We went up to a grove of willow trees above the village and sat down to rest and wait for the others. They were not long in coming. I asked Mrs. Bhavnani about the bridge.

'Wasn't it terrible !' she said. 'I held on to the Tashi Nono, and tried as hard as I could not to look between the planks.'

And that was all. I was impressed, but disappointed that there was no 'story' in her journey across the bridge.

While we were eating lunch a group of women came up from the village and stood around us, grinning and asking for *bakhshish*. Most of them wore a single turquoise at the parting of the hair, indicating a state of spinsterhood. Several of them were well past the marriageable age, and would never get the chance of removing their ornament. We joined in their free and aimless laughter, photographed them, and gave them a little money. They continued to giggle and watch every movement of ours till we left.

I was given a horse to ride. I had by now mastered the technique of dealing with a Spiti pony and, making myself comfortable on the high, well-padded saddle, allowed my body to rock and sway with his movements. But the horse moved only when he was dragged by the man who had brought him, and this man infuriated me. He had the face of an idiot, and he looked at me with silent, murderous hate in every line of his features. As he pulled at the reins to make the horse follow him, he

M

looked back from time to time, and each time his eyes met mine the measure of his loathing for me increased. His feelings found an answering echo in my own heart, and I was suddenly filled with intense hate and an insane anger. I wanted to strike him ; or spur on my horse, knock him down and trample upon him. When I tried to reason with myself, and sought the cause of this horrible state, my inside began to curdle and churn with impotent rage. I realized at last that this was an encounter between two men who had been deadly enemies in a previous existence, and nothing whatever could be done about it. But I could not bear to be in such close proximity to him. I mistrusted him ; I mistrusted myself ; and when, after twenty minutes, we reached Khure, I dismounted and told the man with impatient gestures to follow me at a safe distance. He scowled and breathed fire, but he understood my meaning and stayed behind. I did not see him again till we reached Sumning.

Hatred is a consuming passion and the experience left me tired and unhappy. According to Buddha it is the fifth fetter, constricting the spirit of those who aspire to attain wisdom. He attributed it to a consciousness of diffidence produced by the delusion of individual existence, and showed the way in which a disciple can destroy malice, ill-will and hatred :

He lets his mind pervade one-quarter of the world with thoughts of love, and so the second, and so the third and so the fourth. And thus the whole wide world, above, below, around and everywhere, does he continue to pervade with heart of Love, far reaching, growing great, and beyond measure.

Just as a mighty trumpeter makes himself heard, and that without difficulty in all the four directions ; even so of all things that have shape and form, there is not one that he passes by, or leaves aside, but regards them all with mind set free and deep-felt love.

He continues to meditate, and in this manner fills his whole being with Love, and in succession with Pity, Sympathy and Equanimity. At the end he is able to cast off the fetter of *Patigha* [1] and reach the end of the third stage on his way to *Arahatship*. [2]

Buddha practised what he preached. This is truer of him than of any other sage or preceptor. There is a beautiful story illustrating his complete freedom from hatred of any kind. Once, during his wanderings, he went to the house of a rich farmer, begging for alms. The farmer was indignant and called him a lazy good-for-nothing, idler a parasite, a hanger-on.

'Look at me,' he stormed, 'I labour and toil and earn my living. I plough the land and work on it. Then only do I gather the fruits of the earth. What do you do but idle and beg ? Why should you claim a share in my harvest ? What have you done to deserve a free meal for yourself and your fellow idlers ?'

Buddha answered : 'My friend, do not misjudge me. I, too, labour and toil, for if you plough the earth, I plough the minds of men and sow the seeds of under-standing. I show them how to destroy ignorance and reap the harvest of wisdom.'

The farmer had no patience with this kind of fatuous

---

[1] *Patigha* — the fetter of hate and malice.
[2] *Arahatship* — the status of a perfected disciple. *Arahat* literally means worthy.

twaddle. He heaped more abuse on the holy mendicant and told him to be gone. Buddha craved leave to
ask one question of the rich farmer :

'Should you offer me a hundred silver coins and put
them in my hand, but I should decline to accept them,
what would happen to the money ? To whom would
it belong ?'

'To me, without a doubt. If the coins are not
accepted they come back to me and belong to me.'

'Then, sir, I do not accept your harsh words and
your abuse. I take my leave of you, but I am sorry
that I shall have one friend the less.'

I remembered this story as I walked through Khure,
and strove to free myself from the unreasoning hate
which the man with the pony had, so strangely, aroused
in every part of my being.

Khure is a delightful village of twenty homes surrounded by fields of barley and green peas. The barley
fields were deep green in colour, and the plants were
sprouting into beards. The peas were in flower, making
large patterns of green and pale pink and maroon.
Round every field grew a border of blue and purple
anemones. Here and there a water-course rippled into
the fields with a soft, rustling sound. A gentle breeze
with an occasional playful gust sent shimmering waves
through the sunlit fields, and filled the air with the perfume of green fields and pea-flowers. It was heavenly
to walk through these fields and breathe the scented air.

At Sumning we camped just below the village, on a
piece of dry level ground, caked and cracked by the
sun into a jig-saw pattern. The man who helped me
to put up my tent had the features of a comic cartoon

and an inane look on his face. His name was Sonum Dorje and he wore a static smile except when he laughed. This he did by closing his eyes, wrinkling up the lower half of his face, and braying like a donkey. Every time I looked at him I laughed, and he laughed back. We took a long time to put up my tent, and had many long and immoderate laughs together. Sonum Dorje did me a world of good, and his photograph is a constant reminder of my hilarious afternoon at Sumning.

The next day's journey was a long and tiring march of sixteen miles to Hansa. The plateau on the right bank of the Spiti has, running across it, several deep ravines, carved out by the action of snow torrents. At each of these we had to climb down a steep path to the bottom, cross the torrent and climb up again. This involved an almost vertical descent of 700 feet or more, and an equally long but far more strenuous ascent back to the plateau. We passed through Morang village, crossed the Gyundi torrent, at the bottom of a deep fissure, and skirted Al and Pangmo. Then came another deep ravine, a long grassy plain called Pildhar; yet another ravine; and again a long flat plain, dry and exhausting; and finally a spring where we sat down to rest, and eat our lunch. Once again a steep descent, this time to the Spiti, across a wooden bridge and climb, climb, climb. Surely, I thought, someone must have forgotten to countermand the order of 'advance', or call a halt. Up through Kioto and on to Hansa, where we arrived at half-past two. The villagers were surprised to learn that we from the plains had travelled all the way from Lara since the morning.

As I lay resting on my bed the usual afternoon wind flapped my tent, and brought masses of dust inside.

Extract from my diary for July 18 :

Had good sleep and woke up at 4.10 A.M. feeling fresh and rested. We were ready at 5.30, but could not make a start for nearly two hours, as some of the mules had wandered far during the night. Shrinagesh, Pal and I left at 7.15, and covered the six miles to Losar in two hours. Crossed the river without any difficulty ; though I had a moment of nervous doubt in midstream, when my pony showed a desire to go back. At 10.5 Pal and I left Losar — Shrinagesh and Bachittar Singh stayed behind to check the *patwaris'* work. We walked at a good pace and were at the top of the hill near Kala Khol just before 12. We stopped to eat our lunch and pushed on. Kala Khol was a little difficult ; my horse lacked both skill and will, but I clung on and managed to stay on the saddle. A mile and a half farther, we came to a luscious green meadow with flowers strewn all over it. Here we rested for an hour and watched the whole caravan of our mules go past. We followed, and in less than a quarter of an hour reached the camping-ground on the bank of the Takshi stream.

What a glorious spot this is, and what a pity we are not spending more than a few hours here. I could stay a whole week, reading, writing and sunbathing in these colourful surroundings. I had the tents put up and took a cold bath in the stream. The path to Baralacha takes off at this spot. Forest Officer Sethi who had just come over the pass from Lahoul, met us and camped with us. He told us that the path between Chandra Tal and Suraj Tal is very bad. Three or four glaciers have to be crossed, and the torrent, Topo Gogma (higher stream), is very fierce. From Chandra Tal to Topo Gogma the track runs along a precipitous moraine and is dangerous. Sethi is an active young man and has a

cheerful way of talking about things. He does not seem the
sort of person who exaggerates. Nevertheless, why cross
torrents and glaciers till we come to them? To-morrow,
at least, we have an easy day — an eight-mile stroll to
Chandra Tal. . . .

It is cold in the late afternoon. At 6.30 P.M. the temperature is 45° F. We are nearly 14,000 feet above sea-level.

VIII

THE MOON LAKE

THE next day's march was one glorious picnic. We had a late breakfast, and started at half-past eight. After crossing the Takshi stream the path climbed steeply up to the Balamo pass (15,000 feet). We reached the top at a quarter to ten. It was a bright and clear morning, and the view on every side was magnificent. All around us were reddish-brown mountains, streaked with white snow and topped by snowy peaks of all sizes and shapes, extending in successive ranges as far as the eye could see. Here and there the blues and greys of glaciers glistened and shone in bright patches. An occasional cloud moved slowly across the deep-blue vault of the sky like a giant puff of luminous smoke. At the bottom of the valley, two thousand feet below us, the Chandra torrent lay like a silver ribbon coiled about the foot of the mountain.

The down-hill journey was easy going, and after a short descent we came to a large meadow where some shepherds were camping. There were thousands of sheep grazing on the meadow and the slopes beyond, and the ground was covered thick with their droppings. The shepherds were from Kangra, and we were able to talk to them in Punjabi. They gave us a demonstration of lighting fire with steel and flint. The steel, consisting of a thick blade of iron, is struck against a piece of hard stone, held in the left hand. The spark, flying from the stone, catches a ball of fluff made from dried edelweiss flowers which is held against the stone, and at once the shepherd begins to blow hard on the ball of fluff. As soon as the ball ignites, the flame is transferred to a handful of hay and faggots. It took no longer than lighting a fire with a match-stick, and the skill with which the shepherds used their primitive apparatus was impressive and convincing.

We crossed two ravines, near one of which Pal found a large prism of natural crystal ; traversed a long stretch of stony ground ; crossed a tiny stream, and climbed up a soft grassy slope which brought us to Chandra Tal — the Moon Lake.

I have never seen anything so beautiful and fascinating as the glacier opposite the Moon Lake. It came sweeping down from one end of a snow-spotted ridge, on either side of which stood tall snow-covered peaks of a dazzling white, glistening in the midday sun. This was one of the two branches of the Sumandari glacier, ten miles in length with a mouth two miles wide. The frozen river ran in a wide curve to the right and then to the left, describing a huge inverted S ; and as it proceeded on its way, it changed its colour from the light transparent blue of sheet-glass to a soft green and light grey. In its lower reaches it turned a dark grey and then suddenly disappeared at the edge of a deep-green plain of grass which fell and rose in a mighty billow, before dropping to the river-bank in a steep and sudden cascade. On our side of the valley the Chandra Tal lay in the centre of a huge bowl of grassy downs, covered with a closely woven pattern of edelweiss and buttercups and forget-me-nots, rippling and shimmering with every gust of wind. Beyond the lake, rocky crags rose to a height of 2000 feet, and from all sides snow-covered peaks looked down at us.

The lake itself is a marvel of loveliness. It is three-quarters of a mile long and three furlongs in breadth. Its colour varies between deep green and greenish blue. As we came up to it from the south it looked bright green, like a shining turquoise placed on a cloth of green baize. From where we camped at the northern end of the grassy plain it was deep greenish blue — the colour of lapis lazuli.

After tea we strolled down to it. A cool breeze was blowing across the meadow, and the feel of soft springy sward underfoot was delicious. A few horses stood grazing at the far end, in groups of twos and threes. The velvety petals of edelweiss glistened silvery in the slanting rays of the evening sun. Here and there a buttercup caught the light, and as we went past it suddenly glowed and went out like a tiny flame of gold. The surface of the lake was calm, except for a few soft ripples which came gliding up to the shore and shook the tiny pebbles lying on the beach. The water was clear and of the lightest green colour, and with each tiny wave it moved and sparkled, looking more transparent than colourless liquid. There was a quality of crystalline brightness about it, as if some fairy hand had dyed the entire lake with the juice of crushed emeralds. From the far end of the lake rose a mountain of grass and stone and snow. In the light of the setting sun the grass had a pale and jaundiced appearance, the stone had turned scarlet and purple, and the snow on the peaks above had the flaming colour of nuggets of pure gold. The Sumandari glacier across the Chandra had lost its brilliant colours, and its features were merged in the dark mass of mountains, outlined against the western sky. As the evening light gradually faded, a strange quiet fell on the landscape like a dark and heavy mantle, disturbing the mind and pricking into consciousness feelings and desires that had lain quiescent for days and weeks.

There is an Arabic proverb which says : Three things there be which ease the heart from sorrow — water, green grass and the beauty of women. There was water in the Moon Lake, and green grass all around, but the beauty of women was wanting. I do not know if it was the rarefied air of Chandra Tal which dissolved the resistance of my censor, or if the impact of so much loveliness had released the springs of sensuality ; perhaps it was nothing more than the feeling of continuous physical fatigue which makes men long for warmth and softness, or merely the long period of enforced continence which brought libidinous thoughts crowding into my brain ; but my inside was shaken by unrest, and for a long time that night I lay awake with my arms and breast aching and yearning for something soft and warm to hold close to me ; and when I slept my dreams were made from the stuff of St. Anthony's visions. The spell of the Moon Lake was sweet and rapturous, but it induced a craving which could not be satisfied and left a sense of bitter frustration.

There were others with whom the charm of the Moon Lake played havoc during the night. Early in the morning I heard Mrs. Bhavnani calling out to her husband in the next tent : 'Darling . . . darling.'

There was no reply.

'Darling, I am not feeling well.'

'What shall we do then ?' her husband asked.

'I am feeling very weak.'

'What do you propose to do ?' he asked again. 'We have to go on.'

'I had a very bad night.' Her voice was the moan of a wounded animal.

'I am coming, darling.'

They discussed the whole matter, question and answer, back and forth : Bhavnani showing concern and helplessness, his wife relating the story of a night of wakefulness, headache and disorder of the stomach. She had tried to seek relief by vomiting, but though her inside churned and stormed, it would not allow the contents to escape. Bhavnani ended by repeating that we could not stop at Chandra Tal, and had to go on somehow. I went across to her tent and said a few kind words, and suggested a nip of neat brandy. Bhavnani agreed with me and maintained that it was a sovereign remedy for all stomach troubles as well as for sea-, air- and mountain-sickness. We were not mistaken, and in half an hour Mrs. Bhavnani was up and dressed, ready for the day's toil.

There had been a little rain during the night, and our tents were wet. Our stock of wood fuel had run low, and Chanchlu spent a whole hour looking for twigs and sticks to light the kitchen fire ; but there was nothing in sight except green grass and flowers, and beyond these, barren rocks ; so in the end he extracted

a few tent-pegs from the sack of spare camp equipment
and burned them.   But it was a glorious morning,

> Kissing with golden face the meadows green,
> Gilding pale streams with heavenly alchemy,

and as I looked down at the Moon Lake in its setting of
a huge bowl of green grass I knew that the spell of the
previous night had lifted.

The day's march was hot and arduous, despite the
short distance we covered. We had intended to leave
early, cross the Topo Yogma, and camp a mile or so
short of Topo Gogma which, according to our information, was a fierce and dangerous torrent. This would
place us in a position of vantage, and we should have
the most favourable conditions possible for dealing with
the bogy-river. But our start from Chandra Tal was
delayed. It was not the wet tents that held us back,
but a lamb, the most tender and succulent I have ever
set my teeth into.

We had had no fresh meat for several days, and
tinned sausages were beginning to taste very tinny and
very unlike sausages ; so, when Chanchlu placed his
hands on a baby lamb, pinched it all over and began to
bargain with one of the shepherds we encouraged him.
The result was an all-night cooking orgy. Raw meat
would not have travelled at all well in the intense heat
of the day, and we could not afford to waste any part
of the lamb. There was only one fire and that, too, not
a mighty one. So it was eight in the morning before
we were able to leave.

Shrinagesh was angry, but the dish of liver and
onions, with potato chips added, which Chanchlu sent up
was fit for a king's breakfast ; and when the marmalade

and coffee stage came we found ourselves talking
of summer resorts and luxury hotels all along the
Chandra valley. Bhavnani, the Director of Films Division in the Ministry of Information and Broadcasting,
and the creator of some of the best documentary films
about India, saw possibilities. We might be able to run
a regular air or helicopter service to the bank of Chandra
Tal. In summer there would be hiking, boating and
swimming for the visitors, and in winter, skating and
the finest ski-ing you could get anywhere in the world.
A conducted tour to the Sumandari glacier would be a
most attractive way of spending a short holiday.

We all sighed in anticipation. But when would this
dream be realized ? Not for fifty years !

The path we followed was narrow and rugged. For
its greater part, it lay along the face of a steep precipice
falling almost perpendicularly into the Chandra river.
We had to cross several ravines and fields, covered with
boulders, some of which were so large that they completely obstructed the view, making it difficult to see
where one had to go. We picked our way through
these labyrinths, scrambled down and toiled up the
steep walls of ravines, panting and sweating ; and at
the end of five hours arrived at Topo Yogma with legs
scarcely able to move, and minds sapped by heat and
exertion. We had travelled only eight miles, but the
continuous labour of climbing up and down steep
paths was not calculated to raise our spirits and fill our
exhausted bodies with energy. Mrs. Bhavnani and
Bachittar Singh were definitely 'off colour', and complained of headache. Shrinagesh showed his impatience
with cooks who were forgetful of the passage of time,

and I was teased and tormented by the memory of my home in Simla, full of cushioned chairs and spring-beds and children's voices. On the narrow ledge, perched above the Chandra torrent, the sole significance of existence had been that each time I dragged my foot forward it meant just one step less out of the total number allotted to me by fate.

We made several attempts to cross Topo Yogma, but the torrent was too deep and too fierce. The sun had been working on the snows for several long hours and the *topo* had reached a state of intransigence. We decided to camp on a shelf above the torrent.

My diary for July 20 contains the following entry :

From where we are camping there is a lovely view of two tall peaks. The taller one is a sharp-pointed pyramid which, according to our maps, is 20,000 feet high. The Chandra and Topo Yogma are roaring below us ; and there is a continuous rumble, the groaning of stones and boulders being pushed forward by the torrent. A small patch of dirty snow lies at the back of our tents where the shadow of overhanging rocks stays longest. The sun is hot, and though we are at a height of 15,000 feet above sea-level, I am sitting in my shirt sleeves with the tent flaps open as I write this. I have no doubt the night will be cold.

I am feeling the height since yesterday. There is no headache, and my pulse rate is 82, but the slightest exertion makes me gasp and pant. The process of dressing in a semi-erect position, inside the tent, has to be eased by resting in between the struggles with the various garments, shirt, trousers, jersey, wind jacket. This morning when I packed my bedding, I had to do it in three distinct movements with pauses in between — (i) roll it up and sit on it, (ii) pull one strap through the buckle and fasten it, (iii) pull the second strap and fasten it.

To-day's walk became a painful ordeal towards the end. The way was rough and stony, and crossing the ravines was an agonizing business. I felt tired and limp. I mounted a pony, but I felt insecure in the saddle and dismounted after a few yards. I kept putting one foot in front of the other, stumbling on and on, even my arms felt inert and lifeless, as they dangled involuntarily from the shoulders. If I took two or three quick steps, I had to stop and take breath. Not a pleasant experience — this feeling of being drugged and then being forced to march on and on.

The night was not particularly cold, and when I had put on my full-sleeved jersey with the polo collar, and wrapped myself in the ample folds of my Kashmir dressing-gown, I only needed a *gudma* [1] and two *lohis* [2] for cover. I had a *lohi* to spare. I slept soundly, and a slight drizzle which fell during the night did not wake me.

We were up at half-past four and ready before five o'clock ; but once again we were delayed by the mules having strayed far in search of grazing. We made a start at seven, and though the current of Topo Yogma had abated during the night, it took us more than an hour and a half to push the whole caravan across. We knew we should be too late for Topo Gogma, and resigned ourselves to a short march of five miles. On arriving at the broad plateau which lies above the torrent, we made a gallant though half-hearted attempt to continue our journey to the snow-bridge two miles up-stream ; but after a hundred yards of desperate

[1] A kind of hand-made Kulu rug, thick, soft and fluffy on one side. It is twelve feet long and four and a half feet broad. Folded double, it is warmer than a thick quilt.

[2] A thick woollen shawl, frequently used as a blanket for bed covering.

struggle we came back to the camping-ground. Even Shrinagesh was forced to admit that no horse or mule could tread this path.

But there were compensations. The camping-ground was a delightful grassy plain, with tiny streams flowing through it. We had the most glorious view of the mountain range across the Chandra; and for a long time I sat leaning against a stone, looking at the panorama of snow-covered peaks and glaciers through a pair of binoculars. Directly opposite us stood a huge triangular peak with the snow lying thick and smooth on its sloping surface. Half-way down crevasses were forming like deep gashes over the face of the pyramid; and at the base a great mass of snow, beetling over the lower slopes, marked the starting-point of avalanches. The serrated edge had a fresh appearance, as if a blunt knife had, only a little while before, cut a chunk off the fifty-foot-thick layer of snow and ice. I watched and waited a long time hoping to see a piece of this massive crust break off and go hurtling down to the river-bank, gathering speed and volume; but the pyramid of snow remained unmoved in its static and unflinching majesty.

It had been a cloudy morning, but soon after our arrival at the camping-ground a sharp cold wind arose and cleared the sky. The sun stood almost overhead, encircled by a rainbow of marvellous beauty. This was the second time I had seen a spectacle of this kind; and on each occasion the rainbow lay round the midday sun, and stood out bright and sharply defined against a clear sky. In size it equalled the ring one sees round the moon on a hazy night; but the circle was not quite complete, and about a fifth of the circumference was

N

missing. All the colours of the spectrum were there lying packed together in concentric arcs. The sight of so much beauty made one feel that the blood and toil expended in coming to the Chandra valley were amply repaid by the strange and breath-taking loveliness of this phenomenon.

Early the next morning, I heard Shrinagesh interrogating the muleteers. *Had the level of water in Topo Gogma fallen during the night? What was the depth of the current? Was it safe to make the crossing?* It had rained a little during the night and the sky was overcast with clouds. This meant the snow had not ceased to melt; but the mule-men thought there was a reasonably good chance of fording the torrent.

The Spitials from Hansa and Losar who had accompanied us to show us the way and help us find the places where the two *topos* could be forded were not at all helpful. They rubbed their chins and cast sidelong glances at the far hills. They seemed unwilling to go down and attempt a crossing. At half-past six we sent the whole herd of mules down the precipitous path to the bottom of the ravine, 500 feet below the camping-ground, and watched them make tentative sorties into the water. For nearly an hour the animals marched up and down the bank of the stream, stopping every now and again to sniff at the current and dip a leg, but they were unable to find a safe spot. It looked as if we should have to go all the way back to the Kunzum pass, or stay till the autumn frosts came and made the stream fordable by freezing its source.

We heard Shrinagesh calling to us from where he

stood half way down the steep slope, and saying that
everything was all right as one of the police constables
had succeeded in crossing the *topo*. We hurried down
the cliff, slipping and slithering on loose earth and
stones, and walked up to the spot from where the
constable had started on his heroic journey. The *topo*,
at this place, was a seething mass of greenish-blue liquid,
foaming and lashing itself against the rocks lying in its
path. The roaring and groaning of the torrent filled
the narrow valley with a deafening clamour, but the
red-turbaned constable stood on the opposite bank
thirty yards away like a beacon of hope and courage.

The mule-men led the animals to the water's edge
and began to drag them in one by one. The animals
hung back, strained at the ropes, took one step forward,
stopped and neighed loudly. The men shouted at the
animals and at each other. We had brought ropes with
us which we had not so far used. We took them out
now and gave them to the mule-men. The mule-men
looked at the thick coils in contempt and shrugged
their shoulders — someone had to brave this death-trap
before the rope could be stretched across the torrent,
and used as a support and a safety device. Who would
that someone be ? Why had not the constable carried
one end with him ? The men from Spiti stood on the
bank and looked on in silence. Time passed.

Step by step, moving in slow motion, the animals
were going forward. The mule-men stood knee-deep
in ice cold water and drove them on. Two mules
passed and reached the other bank, a rope was stretched
across the stream, and the men held on to it. One of
the animals got into difficulties, stumbled and fell. He

wriggled and struggled while half a dozen men pulled and pushed to raise him to his feet. The load was taken off and the mule finally stood up. A number of men formed a chain and began to go across. One of them fell and remained for a moment fighting against the current; but his grasp on the hands of the men on either side of him held, and he was on his feet again and struggling forward, soaked and shivering. This decided us: we would cross over by the snow-bridge.

We waited to see the last mule go across, and then turned to climb up the steep cliff. Toiling up the almost vertical slope, finding a foothold on loose earth, or on sharp stones and large smooth boulders, was a penance which mortified both flesh and spirit. Every step set the lungs bellowing, and the heart beating like a sledge-hammer. I had read of pulses bounding in books of romantic fiction; I had a taste of this experience now. I felt the quick throb of blood all the way down my arm. I could count the beats without putting my fingers on the wrist or in the crook of my elbow. We went on because there was nothing else to do, not because there was any wish or will to go anywhere; even the desire to live was dead.

The path, if it could be called that, was rapidly getting worse. We came to a point where there was nothing but solid rock above us and below. But it looked more terrifying than it really was, and we were able to slide down on our hands, feet and buttocks. The snow-bridge lay below us. It was a solid structure of ice and snow twenty feet thick. The Topo Gogma ran out from below it with a loud and angry hiss and went tearing through the gorge beyond on its noisy course.

Climbing up the moraine on the opposite side was comparatively easy. Shrinagesh and I were ahead of the others, and, half way up, we looked back to see how the Bhavnanis were faring. Shrinagesh had told the men from Spiti to guide and help them over the path to the snow-bridge, and far away on the other side of the ravine we saw a line of tiny dots crawling along the hillside, inch by inch, battling with rocks and stones. We had no doubts or fears about their safety. The Spitial may hesitate to ford an unfamiliar stream, but he is not afraid of tackling any kind of mountain-path. His foothold is firmer and surer than a goat's ; and he can carry a load over quite impossible routes without faltering or stumbling. We knew that our Spiti guides would bring the Bhavnanis through safely, carrying them down to the snow-bridge if necessary. Shrinagesh sent back two more men to give added confidence and help.

Shrinagesh and I watched the tiny dots till they had moved beyond the difficult portion of the route, and resumed our climb. Very soon we were on the plateau opposite the camping-ground, and on the path to Baralacha. Here our riding-ponies were waiting for us. We decided to walk on, and told the pony-men to go back and pick up the Bhavnanis. The man carrying our lunch accompanied us.

The path from now on was easier and went up a gentle gradient. Long grassy stretches, watered by tiny streams, alternated with stony patches. Shrinagesh left me and shot ahead whistling to himself. Pal came up riding a pony and passed on. Bachittar Singh joined me, and we shared a pony for some part of the way. We walked on and on till we reached the top of the pass — at least we thought it was the top, because the path had ceased to climb and we seemed to be on a broad saddle. Behind us the path fell away in an easy gradient ; on either side of us mountains sloped up to a great height ; and in front of us we saw a grassy plain stretching as far as the eye could see.

The snow-covered peaks and glaciers of the Chandra valley stood around in a glorious amphitheatre to our south ; the ground underfoot was soft and covered with a profusion of yellow forget-me-nots and butter-cups and a kind of plant I had not seen before and could not recognize. It grew upright from the ground like a tiny bottle-brush with green spikes for leaves, and a red or a yellow button on top for flower. A bright circular rainbow lay round the midday sun.

We were on the pass, but there were five long miles between us and the far end where the descent into the Lahoul valley began. A fine drizzle began to fall. We crossed the Chandra not much more than a mile from its source, went across a meadow full of flowers, crossed another stream, the Yunan, which flows north-wards from the Baralacha watershed to become a tributary of the Indus, climbed up a small hillock — it was raining now — skirted a hill and glided down a steep bank of snow, and found ourselves enveloped in mist.

There was no doubt we had lost our way. Bachittar Singh had, once before, been to the pass ; but on that occasion he had come up from the Lahoul valley, and either he had not come so far or had followed a different route. Be that as it may, he confessed he could not recognize any familiar landmarks. Nirbhu, the *tahsil* peon who was carrying our lunch, did not know the way ; Paldhar, the pony-man, had complained of head-ache and sickness, and I had told him to mount his animal and go on. He knew the way, but he had been gone nearly an hour. Shrinagesh and Pal must be miles ahead and the Bhavnanis were far behind. They had been out of sight since we saw them crawling down to the snow-bridge over Topo Gogma. But they had several men to guide them and bring them safely across the pass. Bachittar Singh, Nirbhu and I found ourselves isolated and lost in a strange world of snow and grass and mist.

It was not a pleasant feeling, being on top of the Baralacha without any bearings. The pass is a long and rambling neck of high land, connecting the central mountains with the main Himalayan range. It marks the cross-roads of the routes to Spiti and Leh — the routes are so little used that there are no visible paths or tracks anywhere on the pass. Its name (*Para la rtsé*, in Tibetan) means 'pass with cross-roads on the summit'. We might find ourselves going northwards towards Leh, instead of westwards towards Lahoul, or we might just go round and round, floundering ineffectively in a maze of hillocks and rocky mounds.

The mist lifted and, ten yards away, we saw fresh hoof-marks. We followed the trail to the edge of the snow-bank, and continued over the grass beyond in the same direction till we came to a spot which Bachittar Singh recognized as the beginning of the

pass from the Lahoul end.   A few minutes later we were
on an eight-foot bridle-path, prepared, levelled and
surfaced by the Public Works Department.   Then came
a mile-stone : PATSEO 11 MILES.   We were back in the
civilized world of mile-stones and furlong-stones.

IX

LAHOUL

THE world of mile-stones and furlong-stones was also
the world of rest-houses and human habitation. Henceforth at the end of each day's march we should come to
a government bungalow with its roof of corrugated
iron painted a bright red, to give cheer and hope to
travellers over the last three or four miles of their
journey. Inside the bungalow there would be beds
and a fireplace, cane-bottomed chairs with long exten-
sible arms to act as foot-rests, and a kitchen where
Chanchlu would be able to light the fire quickly.
There would be hot baths and space to stand erect and
push one's arms straight into the shirt-sleeves till the
hands were free of the cuffs. There would be a sense
of security and comfort, a sense of having accomplished
the assigned task, of having, at last, turned into the
straight run for home.

The P.W.D. path forming the highway to Leh and
Tibet was broad and the gradient varied between
straight and gentle slopes and snatches of steep zigzagging runs. The scenery was rugged and grand.
The rain stopped soon after we left the saddle of Baralacha and we had a clear view of the colourful rocks
around us and, in the far distance, of a high ridge running across the line of sight, with three magnificent

glaciers perched upon it in a row, like china figures decorating a shelf. We passed along the edge of Suraj Tal — the Sun Lake — from which the Bhaga river flows out, scarcely pausing to take in the beauty of its limpid blue water in a setting of grey rocks and stones, and marched on, descending continuously. There were a few tiny streams to cross and an occasional snow-bridge which presented no horrors or difficulties ; we forded a small torrent, passed over a cantilever bridge spanning the Bhaga and·came to the camping-ground of Zing-zingbar at a height of 14,000 feet above sea-level. There used to be a P.W.D. *serai* (inn) at this place where shelter for the night and some provisions were available. The *serai* was burnt down some years ago, and we saw only bare stone walls standing in a picturesque grassy plain which lay well above the Bhaga river.[1] The valley had now opened out considerably and there was an atmosphere of peace and restful ease about the place. A party of shepherds was camping behind the ruins of the old *serai* and I wanted to join them in their meal of thick coarse *chappatis* and potato gruel.

Pal was sitting on a stone by the roadside and con-templating the scene. He told us that Shrinagesh had gone on to Patseo at break-leg speed and there was no possibility of overtaking him. His clothes were wet, and, as there was a cold, piercing gale blowing, he had decided to go on and not wait for us or his food. Once or twice we had caught a glimpse of him striding on, alone and grim-looking ; we had called out to him as loudly as we could, but either our voices did not reach

---

[1] The inn has since been rebuilt and travellers may halt there, but no provisions are available.

his ear, or he paid no heed to them, for he showed no
sign of having received any impact from the external
world and continued to stride along relentlessly without
stopping or looking back. I was quite sure he was
singing his favourite song about the lady who was fresh
as a daisy. It was twenty minutes to five and I had eaten
nothing since breakfast at 6 A.M. The fiend of hunger
was at mine elbow. We were only five miles from
home and I readily agreed to Bachittar Singh's suggestion to sit down near a fresh-water spring and have our
lunch. I told my conscience that Shrinagesh would
soon be reaching Patseo, and at the rest-house he would
find sufficient sustenance and refreshment.

With a good meal inside, the remaining five miles
to Patseo were like lolling in an armchair and eating
bananas. I marched at a brisk pace reciting Gilbert's
verses and Shakespeare's sonnets, stopping occasionally
to pick a flower or a stone of unusual colouring. I was
able to collect a beautiful blue poppy, but, alas, it
suffered a great deal in pressing, and now it is but a
poor caricature of its fresh and brilliant loveliness.

The Patseo rest-house is situated on a grassy plateau
lying high above the river, and commands a delightful
view of the entire valley with its snow-covered peaks
and a glacier on the southern side, and of the deep U-
shaped trough of the valley winding away towards the
north with the footpath clinging to its side like a thin
long serpent. All around the rest-house our tents and
chholdaris had been pitched, and there was a great deal
of feverish activity which did not seem to have any
motive or purpose. I soon realized that it was nothing
more than the exuberance provoked by a return to

surroundings which were more familiar, or at any rate more hospitable, than those we had been accustomed to for several days.

The rest-house consisted of two rooms and two bathrooms with a verandah in front. The kitchen and servants' quarters were at the back. The furniture was simple but adequate. Crockery, cutlery and cooking utensils were available for six. This, indeed, is the general pattern of rest-houses in Lahoul, and though the accommodation is small, there is ample comfort inside the strong stone walls and corrugated iron roofs, lined with a ceiling of pine-wood planks. One room was reserved for the Bhavnanis and the other for Shrinagesh and myself; for the rest, tents were put up. I found Shrinagesh lying in bed, tired and uncommunicative. He had had no food since breakfast, and on arriving at Patseo he had ordered tea and then gone to bed with a whole potful inside him. I asked him to get up and eat some dinner and spoke to him at great length on the subject of his obligations to his own person. He had walked all the way to Patseo and in record time; he had won the admiration of the entire party by a demonstration of courage and fortitude, but I pointed out, quoting Christopher Fry :

> What after all
> Is a halo ? It's only one more thing to keep clean.

I finally persuaded him to have a large dose of brandy and some food.

I had a hot bath and sat down to write my diary and wait for the Bhavnanis. At half-past seven I sent a man with a Petromax lamp to meet them and light their way home. 'Now, don't go round the corner and sit down', I admonished him. The Bhavnanis arrived at half-past eight. We sat talking and sipping brandy till a quarter-past nine when dinner was served. There were only a few bones left of the lamb which had delayed us at Chandra Tal, and with these Chanchlu had sought to add a certain measure of variety and sapidity to his favourite dish of hotch-potch.

At half-past four the next morning Shrinagesh's orderly came and announced the time. Shrinagesh acknowledged the information, turned his face to the wall and began to snore softly. At five o'clock, tea was served. We left our beds reluctantly and after a large and sustaining breakfast Shrinagesh and I set off at a quarter to seven and went down to the cantilever bridge which was constructed on the side of the old stone bridge many years ago (Patseo means literally 'stone bridge'). We had a long march of twenty-four miles in front of us, but the road was good and it was a beautiful morning. The hillside was covered with artemisia. This silvery-green bush gives forth a heavy scent, not unlike that of lavender, and, as the morning advanced, the still, warm air of the valley was filled with its perfume. Artemisia has great medicinal value, for from it is extracted santonin, the specific remedy for worms. We learnt that experiments made with Lahoul artemisia had revealed a satisfactory percentage of santonin content, and steps were being taken to manufacture this drug in India on a commercial scale.

We passed the little smiling village of Darcha, lying almost at the river's edge and surrounded by fields of green barley and a thick grove of willow trees. The path made a wide sweep to cross the Darcha stream over a bridge and arrived at the Jispa rest-house, situated on the edge of a pine and juniper forest. Here we found Thakur Pratap Chand waiting for us with a large bucket, full of snow trout. The trout was obtained, not by the traditional method of tickling it, but by a much simpler device. Two miles from Jispa a snow-stream falls into the Bhaga river ; the fisherman takes a basket down to the water's edge near the confluence of the streams and throws a small stone into the river. The fish rush up the snow-stream and are picked up in the basket. One stone is enough to provide a bucketful.

While Shrinagesh was talking business with the *tahsildar* and Thakur Pratap Chand, I asked for a frying-pan and some *ghee*.[1] These were quickly produced, and in a few moments I had a pair of trout sizzling on the kitchen fire and filling the room with an appetizing smell. When I say that snow trout is far more delicious than ordinary trout, I have said enough to recommend it to the most exacting gourmet. I fried two more and then two more ; but it was getting late ; Kyelang was still a long way away and Thakur Pratap Chand had invited us to lunch at his house in Kolong. Pal and Bachittar Singh had arrived, but the Bhavnanis were reported to be a long way behind. We could not afford to wait for them much longer, so we left, after giving detailed instructions regarding the disposal of the remaining fish. Some were to be fried and served to the Bhavnanis as soon as they arrived and the rest were to be sent to Kyelang to await our arrival. The Bhavnanis were to be asked to follow us to the Thakur's house. But none of this happened. The Bhavnanis were late

---

[1] Clarified butter

in arriving. No one gave them any trout — indeed the entire bucketful of fish disappeared in a most mysterious way. We could not find out what happened to it, and the entire hierarchy of the Commissioner's entourage — *tahsildar*, *patwari*, peons, the rest-house *chowkidars*—presented an impenetrable barrier of stolid silence and wondering ignorance to all our enquiries. I was really sorry, for the trout had a delicious taste and the hurried nibbling at the rest-house had given me a desire for more. The Bhavnanis were not told that they were expected for lunch at Kolong, and after a little rest and their meal of cold *parathas*, they continued on their weary way.

These disappointments, however, came later in the day, and when Thakur Pratap Chand, Pal and I left Jispa we were in a state of pleasurable anticipation, and made light of the hot and tiring walk to the Thakur's house. He regaled us with stories about himself and about the people of the valley. He told us of arrows embedded in solid rock which bore witness to the sheer strength of old-time hunters. He spoke of quaint marriage customs, his own exploits in the army and local legends. He possessed a large measure of the story-teller's art, and his ceaseless recital kept our minds from dwelling on the trials of the steep thousand-foot climb to his house.

Thakur Pratap Chand is a descendant of the old Rajput rulers of Lahoul. His father, Thakur Amar Chand, assisted the British Government during World War I and took more than a hundred men to Mesopotamia in the 6th Labour Corps. He held the temporary rank of *jemadar* and was awarded the title of Rai Bahadur. He was afterwards invested with the powers of an Honorary Magistrate and Civil Judge. Thakur Pratap Chand served in the Indian Army for sixteen years and held the rank of Captain. He resigned his commission in 1948 to manage his estate more profitably and to maintain his personal touch with the people of Lahoul, for he likes to feel that the inhabitants of the valley are still his subjects. Two of his younger brothers are also in the army and hold the rank of Lieutenant-Colonel. The eldest brother of the family, Thakur Abhe Chand, has, for many years, been somewhat eccentric and indifferent to his surroundings. On more than one occasion he has invoked the skill of psychopathic experts but without appreciable results. We saw the old boy in the family home and were introduced to him. He sat with his legs crossed and his back resting against the wall — a picture of quiet contentment and self-sufficiency. From time to time he looked at us and laughed silently, but his merriment asked for no reaction from us, and there was no attempt at any kind of communication with those present in the room. Thakur Pratap Chand told us that he was never violent and lived a life of harmless indolence, washing and shaving when the fancy took him and eating his meals with a regularity which would have done credit to the digestive machinery of a professional athlete. We heard a malicious rumour that Thakur Pratap Chand had, in some mysterious manner, contrived a state of affairs whereby the family property remained in his absolute control, but I have no doubt at all that this accusation was baseless and he was in no way responsible for his brother's misfortune.

The Thakur home is built in the form of a small

castle, well fortified and strongly placed on the top of
an eminence commanding the valley. We entered a
small courtyard, went up a wooden ladder, climbed a
short flight of steps, and, after tracing our way along a
dark and labyrinthine passage, passed through several
rooms of which the floors had been freshly mud-
plastered, and finally found ourselves at the door of a
bright, clean and well-furnished room. This was obviously the apartment where guests of honour are received. We removed our shoes and entered. The
floor was covered with a number of Chinese carpets and
at the far side a low dais ran along the wall, at window
level. Carpets of richer and brighter design were spread
over the dais, and the window provided a magnificent
view of the valley and the distant mountains. We
made ourselves comfortable on the dais and looked
round the room. *Thankas* and framed photographs
decorated the walls ; there were two glazed cupboards,
one containing books and cheap china ornaments, and
the other a number of liquor bottles. In one corner of
the room stood two rifles and a shot-gun. Chinese
lacquered tables occupied handy positions near the dais
and in the centre of the room. Thakur Abhe Chand
sat at one end of the dais smiling blissfully to himself.
We were served with *chhang* and *arak*. The taste of
both beer and whisky was remarkably good and we
drank large quantities of each till we glowed with
warmth and a sense of delightful insouciance. The
arak was delicately flavoured with spices and saffron
and was drawn from the lot specially distilled for the
festivities in connection with the marriage of Thakur
Pratap Chand's daughter which was to take place within
o

a few days. I praised the quality of the liquor and the Thakur made me the present of a bottle. I carried this to Simla and, used sparingly, it served to entertain my guests at several dinner-parties. Poured into small Venetian liqueur glasses with miniature gondolas floating round the base, and passed round as a post-prandial stimulant along with post-prandial anecdotes, the *arak* won general approval and was sometimes preferred to kirsch or Benedictine.

Thakur Pratap Chand showed us the family *sanads* (certificates of merit) and albums containing photographs of himself in various scenes of army life. He had fallen into disfavour with the British because of a suspicion that he was in some way connected with the murder of Dr. Bernardo of the Moravian Mission in 1945. On one occasion the British S.D.O. of Kulu was publicly rude to him and the Thakur lost much of his prestige. He was anxious to be restored to his position of confidence and dignity, and made efforts to please and impress us. The meal he served was a veritable feast, consisting of a large number of dishes excellently cooked. There was *pilau*,[1] flavoured with saffron, mutton-curry with potatoes, fresh mushrooms, turnips, *moong*,[2] *dal* flavoured with mustard, home-made sausages, *sampa* served with butter, sugar and tea, buckwheat pancakes generously buttered, and stewed apricots. There was a copious supply of *chhang* and *arak* to wash all this down and induce a feeling of mellow satiety. We had not eaten such a meal for years and did full justice to the Thakur's hospitality.

---

[1] *Pilau* — rice cooked in butter with meat or vegetables, and spices, raisins, etc.   [2] *Moong* — a kind of pulse.

After lunch we were conducted to the family chapel and shown a number of *thankas*, which we praised, and a small effigy of Buddha enshrined in a glass case. This, our host told us, was one of only two in existence — the other one being in the possession of the Dalai Lama. The figure was made of clay and was not more than two inches in height, but it was possessed of a miraculous feature : its hair grew, slowly but unmistakably. The Buddha was brought from Tibet by Thakur Pratap Chand's father many years ago. At that time the hair was scarcely a quarter of an inch long and fell short of the neck ; now it had grown to nearly an inch and came down to the waist. I examined the hair closely and touched it with my fingers. It looked and felt like real human hair; and grew out of the Buddha's head naturally — each individual hair sprouting from the scalp and joining the others to form a loose tress hanging over the back in a realistic manner. Everyone in the valley has heard of this Buddha and believes that its hair grows. Our host told us that the effigy had brought him luck, and if the hair ceased to grow some terrible catastrophe would strike the family.

We talked to the Thakur about local marriage customs and asked him if matches were arranged by parents and relations, or if the young people chose their life-partners themselves. He told us that both forms were prevalent, though the more orthodox type of marriage for which the negotiations were conducted by the family was far more common. It was, however, not unusual for a young man and a young woman to fix things up on their own, without the consent, and sometimes even without the knowledge of their parents, and elope in the way lovers elope the whole world over. They remain absent for a few days and then return home to present the family with a *fait accompli*. Everything is forgiven, the union is accepted and the couple receives the blessings of all concerned. This, however, is looked upon as a drab and somewhat crude manner of entering the state of matrimony and one which is more suitable for those who occupy the lower strata of society. The orthodox form of marriage has the sanction of the aristocracy and finds greater favour with even the common people. This is a most intriguing and picturesque affair and involves a great deal of complicated ritual.

Negotiations open with the visit of the boy's father and maternal uncle to the house of the chosen bride. A pot of *chhang* is offered to the girl's parents and the object of the visit is explained in ceremonious terms. The mother goes in to ask her daughter if she is agreeable to the proposal. If the girl has other ideas, her parents refuse to drink the *chhang* and there is an end of the whole matter, but if the reply is favourable, the offer of marriage is accepted, and the *chhang* is passed round and drunk in a festive manner. On two further occasions this ceremony of offer and acceptance, accompanied by the drinking of *chhang*, is repeated, and on the third occasion a rupee is paid to the girl's parents and accepted by them. This finally settles the matter ; the betrothal is now complete and a day for the wedding ceremony is fixed.

On the appointed day the bridegroom stays at home while his sword and shield are carried by his relatives and friends in festive procession to the bride's house. Presents of clothes, ornaments, cakes and *chhang* are carried on trays. The bride's people meet this party about a furlong from their house. As may be expected, both parties are in a merry mood induced by the happy occasion and reinforced by the consumption of large quantities of *chhang* and *arak*. A battle of wits now begins. All along the way to the bride's house a line of stones, twenty-one in number, has been erected. At the first stone the bride's party poses a question which must be answered correctly by the bridegroom's party. If the right answer is forthcoming the stone is knocked down and the procession advances to the second stone. Here a second question is put. A wrong answer or delay in replying provokes a great deal of good-humoured jeering and taunting from one side and mock apologies from the other. The ritual is repeated till all the twenty-one stones have been negotiated. The questions and answers, which are all in verse, are out of a book of catechism which is assiduously studied by both parties during the days preceding the wedding ceremony.

At the bride's house the final test of intelligence and valour takes place. A sheep's heart which has been secretly buried must be located and dug up with the bridegroom's sword. Information of the exact spot is conveyed to the groom's party by means of a song, giving the exact measurements and directions. The riddle must be solved and someone from the bridegroom's party must go up to the spot and, with one stroke of the sword, bring up the heart. Attempts, often successful, are made to bribe one of the bride's relations and obtain the secret from him. After this ceremony the bridegroom's party enter the house and display the gifts they have brought with them. Food and drink are passed round. When the feast is over, the bride's dowry is presented. This varies in richness and quantity according to the station and purse of her father. It may consist of several complete outfits of clothing, ornaments, a sum of money, utensils required for the new home, a cow or a yak and a pony or two. The dowry is admired and praised in adequate terms, and the bride is, at last, taken home to the bridegroom, who has, no doubt, been waiting impatiently for her. But he must possess his soul for a little while longer, for the bride cannot enter his house till the party has been cleansed of any hidden evil which they might have picked up on the road. The spirit medium is called. He first invokes the benevolence of the gods, and then going up to the top of the house, throws a live sheep before the wedding party waiting outside. The sheep is quickly seized and torn to pieces ; its heart and liver are cut out, and everyone present scrambles for a piece and eats it raw. All the while the lama reads the sacred texts to scare away the demons. He has with him a small earthen pot with the effigy of a demon, made from dough, inside it. The pot is broken and the effigy is destroyed. This completes the annihilation of the demon and the party enters the house to feast and make merry.

In Lahoul, as elsewhere, there is no fixed age for marriage, but, generally speaking, the richer people marry earlier than the poorer. Marriages contracted in childhood may be dissolved if, on growing up, the spouses find that they cannot agree ; but such separations are rare and are not looked upon with favour. If a marriage proves childless, the husband takes a second wife, but her status is inferior to that of the first wife, for she serves only to produce a child — a task which, but for some slight accident or shortcoming, the first wife could have performed equally well. She does all the outdoor work while the elder wife retains the position of honour and presides over the household. Divorces are not unknown, and a divorced wife takes away her dowry with her. She is also paid a sum of money by way of compensation if the divorce was initiated by the husband ; but if it is she who wants to break the marriage tie, she must pay a fine to her husband. The marriage is dissolved by a somewhat touching ceremony. The spouses take a thin thread of wool, wrap the ends round their little fingers and, holding their hands close to each other, repeat that henceforth they will have no more to do one with the other. They then pull their hands apart, and, when the thread breaks, the divorce is complete. What memories of connubial life and regrets must flash through the minds of the unhappy pair as they look into each other's eyes and pull at the tenuous thread which represents the closest and the most intimate union that human beings are capable of !

In the matter of inheritance the Thakur families follow the rule of primogeniture. The eldest son succeeds to the family estate ; the younger sons, as long as they continue to live in the family house, are maintained by their elder brother, but if they separate they are given a small allotment of land. The rest of the landholders follow the usual agricultural custom of the Punjab whereby all sons inherit equally. The holdings are, however, kept joint, and the sons with their families live in common.

Thakur Pratap Chand told us of a curious custom the Lahoulis have of cheating death. When a sick person cannot be cured by ordinary means, and *vaids* [1] and medicine men have failed to bring him back to health, and his end seems inevitable, an attempt is made, as a last resort, to cheat the angel of death. A life-size effigy of the sick man is made ; it is painted to resemble him, and bedecked with his clothes and ornaments. The effigy is taken out in a mock funeral procession with firing of guns and chanting of hymns by a lama. Fireworks are let off. At the cremation ground the figure is cut up into pieces and burnt in the manner of a dead body. A hired mourner loudly bewails the death of the sick man, calling out his name over and over again and saying that the man has been dead these nine years, so that the angel of death should remain under no misconception regarding the identity of the person whose body is being cremated, and should withdraw his murderous hands from the real object of his attention. Sometimes this device is successful, but sometimes the relatives have to incur the expenses of a second and genuine funeral.

The Lahoul valley is richer and supports a much larger population than Spiti, the area under cultivation in Lahoul being 3445 acres as compared with 2374 acres in Spiti. The population of Lahoul is 10,142 as against the 2422 of Spiti. The annual land revenue of Lahoul is Rs.6900 — which is more than double that of Spiti although the area under cultivation is not quite $1\frac{1}{2}$ times

[1] *Vaid* — physician employing an indigenous system of medicine.

the corresponding area in Spiti. The reason lies in the
greater industry of the Lahoulis, a more fertile soil and
better facilities for irrigating their lands. Before the
Communist influence in Tibet, large quantities of *koot*
— the root of a bush used in the manufacture of incense
— were exported to Tibet and China, and high prices
were fetched for this commodity. This trade has almost
ceased now, and *koot*-growers complain bitterly of
having been impoverished by political changes across
the border. Another complaint relates to the high
prices for *pashmina* [1] wool demanded by Tibetan traders
and the consequent decrease in the import of wool.
This has adversely affected the shawl industry in Kulu.
Potatoes are being grown in increasing quantity, and
the Lahoul potato is now a much-sought-after commodity on account of its larger size and better taste.

The Government of India is, of late, paying special
attention to the development of Lahoul and Spiti. A
jeepable road has been constructed from Khoksar, at
the foot of the Rohtang pass to Zingzingbar, and another
from Khoksar to the Kunzum. The Lahoulis are not
worried about the problem of bringing jeeps across the
Rohtang. This may be a distant and perhaps unattain-
able prospect, but for the present the construction of the
road has brought money into the pockets of the Lahouli
labourers. Another work taken in hand by the Public
Works Department is the construction of a water channel along the face of a sheer rocky precipice, a thousand
feet above the road. The channel has been blasted out
of the vertical rock along a large portion of its length.

[1] Wool made from the under-fur of the Tibetan goat, used in the
manufacture of Cashmere shawls.

It is an engineering feat of considerable magnitude involving a great deal of ingenuity and risk, and, viewed from below, it gives one the impression of a fantastic scheme designed to serve some fairy god or monster residing in the unapproachable region beyond the rocky ramparts. The Lahoulis contemplate this distant wonder with incredulous eyes and shake their heads, but they have learnt to claim large sums of money by way of compensation if any of their trees are damaged by rocks falling from above during blasting operations. Also they know that the water-course means immediate gain in the form of workmen's wages and money spent in the valley by engineers, contractors and imported labour. In the beginning there was some opposition to the project by conservative elements, but, as the work advanced, the Lahoulis began to accept it as a mad but profitable venture.

The Lahouli is, on the whole, a likeable creature. He is a member of a closely knit community and not very friendly or communicative to an outsider, but there is no evil or malice in him. He is not easily aroused unless he is oppressed or feels that his economic security is in jeopardy, or unless his citadel of tradition and orthodoxy is being attacked. The Moravian Mission had a branch stationed at Kyelang for nearly a hundred years, but they were not able to count many converts to Christianity. Their work in the valley was very valuable. They held knitting classes for women, improved the crops, introduced buckwheat into Lahoul ; they opened a dispensary and a school at Kyelang and set up a printing press. Their example was a good influence on the people of the valley and on their living

conditions. But suddenly in 1945 they thought the
Mission was encroaching on their religious and emotional susceptibilities and Dr. Bernardo, a prominent
member of the Mission, was murdered. The Mission
decided to withdraw after a stay of ninety-one years.

Sir James Lyall, who was posted as Sub-Divisional
Officer, Kulu, in 1862, summed up the Lahouli character
in the following terms :

The character of the people is solid and conservative ;
their power of united action is considerable ; they seem to
me not quick-witted, but eminently shrewd and sensible.
Though they show great respect to their hereditary nobles
and headmen, they would, I believe, combine at once to
resist tyranny or infringement of custom on their part. The
headmen have certainly been, hitherto, very careful not to
offend public opinion. Murder, theft or violent assaults are
almost unknown among them, and they seem to me to be
fair and often kind, in their dealings with each other ; on
the other hand . . . the standards of sobriety and chastity
among them are exceptionally low. Drinking is a common
vice in all cold countries, and the want of chastity is accounted
for by the custom of polyandry, which leaves a large pro-
portion of the women unmarried all their lives. In spite of
these two frailties they seem to me to be an eminently religious race ; they seem to think that to withstand these
particular temptations is to be a saint, and that in ordinary
men who do not aim so high, to succumb is quite venial.
The lives of their saints are full of the most austere acts of
virtue and mortification of the flesh commencing from the
cradle, which are certainly calculated to make the ordinary
mortal abandon the task of imitation in despair ; and their
religion, though it fails here, has, in my opinion, considerable
influence for good in their minds in other respects, more at
least than the forms of religion practised by other races. . . .

This is not surprising, as the moral teaching to be found in the Buddhist books is of a very high order. The love of one's neighbour is one of its principles, and this is extended to include even the brute creation. So, again, though good works are balanced against sins, yet their worthlessness, when not done in a humble and reverent spirit, is recognized.

This is a true assessment of the people of this strange and magnificent valley.

We left the Thakur's house at half-past three, and a narrow winding track through his barley fields brought us on to the main route to Kyelang. The path was dry and undulating, but the scenery on all sides was magnificent. On our left stood a high rocky ridge, massive in appearance and running level like a stupendous wall. An occasional glacier rested on a niche near the top of the ridge. On the other side of this wall lay Sisu, two stages beyond Kyelang, but only a few hours' journey for an intrepid mountaineer prepared to scale the precipitous ridge up a rocky route which no goat or sheep would attempt. On the right the hillside went steeply up a thousand feet, rugged and barren, except for a straggling juniper tree standing forlorn in its desolation.

We reached Kyelang at seven o'clock and were greeted by a large crowd, gathered at the rest-house gate, to welcome the Commissioner and garland him. Kyelang is a big village, boasting a population which comprises nearly half the people of the entire valley. It is the headquarters of the Lahoul sub-*tahsil*, and enjoys the privilege of having a post-office and a police-station. There is a government hospital with a Sub-Assistant Surgeon in charge, a high school with over a

hundred pupils, and a co-operative shop where packets
of tea, cigarettes, tins of condensed milk and flour are
available. It is becoming an increasingly important
trade-post and clearing centre for wool and potatoes.
There is a real street in the village (the only one in
Lahoul) with shops and houses, and dirt lying all over
it with an occasional dead rat to add piquancy to the
bouquet of variegated smells presented by the Kyelang
bazaar. There was no doubt at all that we were back
in civilization ; but, mercifully, the rest-house stood at a
safe distance from the smells and squalor of the village,
and we spent a restful evening.

Dinner was late in cooking. I was very tired, and
the vast quantity of food I had consumed at the Thakur's
house made further nourishment for the time being unnecessary. So I turned my back on the dish of steaming
hot hotch-potch and went to bed on a light stomach
and slept soundly for ten hours. Even so I felt stiff and
tired in the morning. Shrinagesh said the best way to
get rid of the stiffness was to pay a visit to the Kardung
monastery perched almost at the very top of the high
mountain wall across the Bhaga river. I looked at the
tiny white dot in the far distance and told Shrinagesh
that a long restful morning in the sun, with a book to
keep me company, would be a more profitable occupation than toiling madly up those slopes to the monastery ; but Shrinagesh had already sent word to the
lamas and he must not fail them, so I allowed myself
to be persuaded to accompany him. We walked to the
village and climbed down the steep path to the Chandra.
Over the bridge and up we went, on and on, half way
to the sky. The exercise loosened my limbs and I found

myself going up the steep mountain with an eagerness I had not anticipated. As soon as we were within sight of the monastery a fanfare of trumpets and drums greeted us, and we saw the lamas standing in a row, on the roof of the monastery. Wearing their ceremonial clothes and silhouetted against the sky, they presented a colourful and imposing tableau. As we approached nearer, the lamas came down to receive us at the monastery gate.

The head lama was a bearded old scoundrel with a pimply nose and a knowing look in his lecherous eyes. His face, at the moment, was in quiet repose, but the lines round his mouth and a glint in his eye spoke of feverish activities and orgies of self-indulgence in which his wicked body must frequently have taken a prominent part. There were several *chomas* (nuns) living in the monastery, and some of the younger ones were not unattractive. Indeed, one of them, who giggled and frisked about provocatively quite unlike a nun, had a great deal of physical charm and I am extremely doubtful if the saying of prayers was her sole occupation in the monastery. I had heard that tantrism was practised in some parts of Lahoul, and I felt sure that the Kardung monastery was one of the strongholds of this strange and interesting cult. Tantrism is a mixture of the old Hindu yoga practices and mysticism superimposed on the Mahayana form of Buddhism. The tantrics emphasize the use of sexual symbolism, and by means of ritual and yoga exercises seek to harness the power of sex for attaining salvation. In its lower forms tantrism is little more than indulgence in perversities and secret demonstration of extraordinary sexual powers; but in its

higher and more advanced phases it is a means of liberat-
ing the mind from the bondage of the ten fetters and of
escaping from the delusion of self and individuality.
Some of the feats performed by tantrics are quite incredible, and I would have liked to witness an exhibition
of tantric powers. But the Kardung lamas were not
disposed to be friendly or communicative.

It was prayer time and we could not see the large
assembly hall where most of the *thankas* and treasures
are stored. We sat in a small room, attended by two or
three monks (the others had joined the assembly at
prayers), and felt rather than heard the low moaning
chant of the lamas creep into the room and fill it with
a sorrowful heaviness aggravated by fumes of incense.
The monks showed us one or two *thankas* and a few
wall-paintings, and then placed before us an appeal for
funds neatly drawn up in elegant Hindi, and mounted
on a square of thick cardboard.

I procured a copy of this document, and give below
a literal translation :

HAIL ! VICTORY TO THE JEWEL
APPEAL FOR HELP

This monastery was founded about 900 years ago ; but it
lay in ruins till 1912. The revered lama Norbu, a resident
of Kardung village, renounced his worldly goods and took
up the life of a mendicant. He travelled on foot to distant
places and visited Bhutan, Tibet, Lhasa, Kham, etc., and
made a deep study of holy books. For some time he lived
with the great Togdan Rimpoche in his cave in Kham, and
practised yoga exercises. He acquired the power of freeing
his soul from his body at will, and attained the status of a
true saint. Then this saint came home, bringing with him

the nectar of wisdom for the good of his people. He preached the gospel of *dharma* throughout Lahoul, and the lamp of his teaching infused new life into this dark and gloomy region. His example and the force of his teaching converted the sons of many families in Kardung; they entered the holy order and became his disciples. In 1912 he renovated the monastery and spent a large amount of money on its repairs.

But to our great sorrow, in 1946, he abandoned his mortal body at the age of sixty-five, and attained the state of *Nirvana*. At the same time it is gratifying to know that to save the ignorant and lead them out of the darkness of doubt, he again took birth in the city of Lhasa. This has been revealed to us through the occult power of lamas and the science of astrologers. We, the lamas of Kardung, fervently desire to recall our head lama incarnate from Tibet and entrust the abbacy of this monastery to him. But we have no funds and this misfortune stands in our way like a mountain that cannot be crossed; for in order to bring out a lama incarnate a large sum of money must be paid to the Government of Tibet. So we appeal to our friends to subscribe generously in cash or in kind, to make it possible for us to solve our problem.

May the blessings of Truth and Happiness rest upon all.

Your petitioners
All the mendicants of Kardung.

The monks seemed impatient to terminate our visit and we felt that our patronage and encouragement were being looked upon as an unwelcome intrusion on their privacy. The mournful drone of the lamas' chant continued to flow into the room ceaselessly. We made a small contribution towards the monastery funds, expressed a pious hope that the incarnation of the head lama would soon be installed in his rightful place, and

got up to go. As we passed by the door of the chapel
we caught a glimpse of the lamas and *chomas* standing
round the altar in their allotted places in attitudes
of complete dejection and repeating their mournful
lament.

Back at Kyelang we found Thakur Pratap Chand
waiting for us. He had been invited to lunch and had
ridden over from Kolong. One of our men had shot
a brace of pheasants, and, in returning the Thakur's
hospitality, we felt that we had not entirely disgraced
ourselves. A free afternoon enabled me to catch up
with my diary and do a little reading. In the sun the
heat was scorching, but in the shade of a willow tree
the air had a quality that stimulated the senses in a
gentle unobtrusive way. I spent two delightful hours
of recuperative sensuality.

The next day's march was an easy one. We had to
go no more than ten and a half miles to Gondla. Nevertheless we made an early start, leaving before seven,
after an early breakfast. Two miles from Kolong lies
the small village of Billing, with its monastery resting
snugly in a small niche carved out of a vertical precipice,
a thousand feet above the village. The path to the
monastery consists of the merest trace along the smooth
face of the rock. I wanted to visit the monastery, for
it had had a special interest for me ever since I saw, in
the house of a friend, a small figure of Buddha which
had been bought from the head lama of Billing many
years ago. The figure was made of bronze with a covering of gold leaf. It was scarcely three inches high, but
the seated Buddha had such a wise and tantalizing smile
that I had long stood spellbound in front of it. I

P

remember that I scarcely heard my friend relating the story of its sale and purchase. I could not make out what the smile signified. One moment it seemed to say : Take your full measure of the joys of the flesh ; there is nothing else so real and nothing that gives such immediate satisfaction. The next moment the face of the Buddha seemed to take on an air of serene austerity and speak the message of *Aniccam, Dukham* and *Anattam*. I had suddenly come back to the awareness of my surroundings, and had been shocked to find myself in a roomful of people talking banalities and laughing inanely. I had spent a most unhappy evening, my mind disturbed by the smile of the bronze Buddha and my senses tingling with the gaiety and loveliness of the company around me.

Shrinagesh and I scrambled up the steep, rocky path to the monastery and arrived panting and expectant at the door of a mean-looking structure of mud and stone. There seemed to be no one at home. We went in. Everything had a shabby and tawdry appearance. The chapel was unswept and unkempt. On the altar lay three clay figures of Buddha, sulking in murky surroundings. The brass lamps on the table in front of the altar had not been polished for months. They were not arranged neatly in a row, nor were they stocked with oil and wicks. One of the lamps lay on its side in a state of helpless disability staring at a bunch of four or five, huddled together at one end of the table. The walls were bare of *thankas* or paintings. The floors had not been mud-plastered for a long time, and instead of the pungent smell of burning incense, there was a dank unhealthy odour in the prayer-room. Everything

looked and smelt of neglect. The head lama lay
bed-ridden in another room. He looked old and sick.
On seeing us he simply shook his head as he had shaken
it when his wife sold the bronze Buddha for a handful
of silver rupees. She had never seen so much wealth
heaped together, the monastery had always been poor,
the desire to lay her hands on the shining mass lying
before her eyes had so overpowered and blinded her
that the lama could do nothing but shake his head in
impotent silence. The bronze god had been handed
over to the rich visitor, and the money had gone to
procure necessities for herself and her husband. But the
lama knew that his wife had done wrong, and, when
misfortune came in the shape of sickness and penury,
he made no effort to fight back. He accepted defeat, for
he knew that this was his due for the sin of acquiescence
in his wife's cupidity.

There was very little we could do for the lama. His
wife had gone down to the village and would be returning in a little while. The villagers of Billing supplied
the bare necessities of the lama and his wife. There
were perhaps half a dozen monks attached to the monastery and they tilled a small area of land with which the
monastery was endowed. They and their *gompa* would
never prosper, but they would continue to exist.
We made a small offering to the sick lama and saw
his face brighten up with a sad smile as we left his
bedside.

Another two miles brought us to the Tandi bridge
and the confluence of the Chandra and the Bhaga. The
two rivers, starting from points close to one another on
the Baralacha watershed, flow in opposite directions,

and after encircling a huge triangular mass of mountains arrive at Tandi from opposite directions and join together to form the Chandra-Bhaga, or Chenab as it is later called. At Tandi there is a magnificent *chorten* to remind the traveller of the danger lurking near the banks of the two rivers and of the necessity of appeasing the gods before crossing, and offering prayers of thanksgiving when safely across. A little higher up, by the roadside, is a *mane* wall of which the top layer has some of the most exquisitely carved stones I have ever seen.

From Tandi the path continued for the remaining six miles to Gondla in a series of undulations, climbing steeply up to the top of a cliff overhanging the Chandra, or descending to cross a beautiful valley forming a by-lane of the main passage through which the Chandra flows. A gentle drizzle which fell continuously all the way kept us cool and eased our toil.

We saw the air-strip which Shrinagesh had proposed for small aircraft. It was a long grassy shelf resting below the bridle-path, half way down to the river-bed. It looked very flat and very picturesque, with sheep grazing all over it, and a few tents massed at one end. *Gaddis* in their passage through the Lahoul valley always camp there, for the grass is rich and plentiful. The strip, I judged, was a hundred yards across and nearly a mile in length, but I doubt if anything but a helicopter can manœuvre safely into the narrow gorge formed by the tall mountains on either side of the valley. The rocky precipices rise to a height of nearly eight thousand feet above the grassy shelf and the distance between them is scarcely more than a mile. The aircraft would have to

climb up to an altitude of 18,000 feet and then swoop
almost straight down to the landing-ground — a quite
impossible feat.

The wind was rising and we quickened our pace.
As we turned a corner a clump of willows appeared,
and soon we were walking through the main street of
Gondla village. It was completely deserted. Beyond
the village, near the entrance to the rest-house, there was
a small crowd of men and children watching our tents
being pitched.

Sitting on the verandah of the rest-house and watching the rain gather force was a cold business, and soon
we had a roaring fire of pinewood logs burning in one
of the rooms. While lunch was being heated I picked
up the visitors' book and began turning over the pages.
The entries went back to 1940 and there were many
familiar names. Former Sub-Divisional Officers and
Deputy Commissioners found mention over and over
again, the Commissioners only occasionally, and the
Punjab Governor and his party once. The entries with
one or two exceptions were brief and not very informative — just a single line for each visitor running along
a number of columns ; name and rank, date and time
of arrival, date and time of departure, whether on
official duty or no, amount paid to the *chowkidar* or
caretaker,[1] and finally remarks. The last column was
not easy and had apparently taxed the literary capacity
and the ingenuity of most visitors. It was usually
left blank. A few were effusive enough to express
their satisfaction by writing the single word 'Thanks' ;

[1] Visitors staying in government rest-houses, unless they are on duty,
pay a small amount by way of rent.

P 2

one or two complained of the *chowkidar*'s incompetence ;
one traveller found him rude and disobliging. The
strong-minded Commissioner who saved the thousand-
year-old deodar trees of Manali by declaring them
ancient monuments paid two visits to the valley. On
the first occasion he was furious because the commode
openings were too small for his unusually large posterior,
and he said so in no uncertain terms. It was during his
stay at Gondla that the famous incident which compelled
him to take disciplinary action against his stenographer
occurred. The story is well known in official circles in
the Punjab and deserves to be related here.

When he went to the bathroom on his first morning in the Gondla rest-house, he saw that the small
circular hole in the commode was wholly incommen-
surable with the amplitude of his person, and quite
apart from involving him in a great deal of discomfort,
would provide too small a target for his activities.
So he banged the bathroom door, and stole out to a
grove of willow trees on the slope behind the rest-
house. He had scarcely arranged himself in the appropriate posture when he caught sight of his stenographer
sitting in an equally inelegant pose behind the trunk of
another willow tree, only a few yards away. His person
was ill concealed by the willow, and the Comissioner
saw that not only was his own person exposed to the rude
gaze of a common stenographer but the horror of the
situation was intensified by the indignity of being compelled to watch a lowly clerk in circumstances wholly
unwarranted by the *Civil Service Regulations* or *The
Government Servants' Conduct Rules*. Immediate escape
was impracticable, and, as the slow moments passed,

the Commissioner's anger mounted till he could scarcely contain it.

As soon as he was bathed and dressed he sent for the miscreant and formally charged him with 'indecently exposing himself to the Commissioner'. After a brief interrogation he found the man guilty, recorded his finding and 'severely reprimanded' him. He ordered that an entry to this effect be made in the stenographer's service-book. The stenographer later made capital out of the incident and showed the entry in the service-book to his friends and colleagues, giving it an interpretation which made them laugh with loud and ribald guffaws and bestow wholly undeserved praise on his physical attributes.

While we were at lunch, Thakur Fateh Chand of Gondla came to pay a courtesy call and invite us all to dinner. We remembered the lavish fare at Thakur Pratap Chand's house and readily accepted the invitation. After some polite conversation we began to yawn rather obviously. As soon as our visitor had taken leave of us, I went to bed and slept like a log till five in the afternoon. It was the most refreshing sleep I have ever had, and when I woke up I felt as if I had been drinking deeply from some strange and hitherto untapped source of energy.

In the evening we went to dine with Thakur Fateh Chand. His house is a seven-storied tower standing in the centre of the village. The lower stories have low ceilings — none of them being more than six or seven feet high, and the total height of the tower is less than fifty feet. The guest rooms are situated on the top storey, and thither we were led up dark and winding staircases to the accompaniment of warnings from our host : 'Mind your head, sir,' 'To the left, Justice Sahib,' etc.

We were introduced to Thakur Fateh Chand's younger brother, Thakur Nirmal Chand, who is employed in government service as Excise Inspector for Lahoul. We found him a pleasant young man, though I could not discover what exactly he was doing in an area where everyone brews and distils his own liquor. We sat down on the floor round a white sheet spread over an old and threadbare Chinese carpet, and were served with *chhang* and *arak*. Both beer and whisky smelt absolutely foul and had a revolting taste. Our host was, however, quite insensitive to these small details and lustily proclaimed that his whisky was better and more potent than anyone else's in the whole valley. He would not take a no, and insisted on our having large measures of it poured out into glass tumblers which were foggy with dirt and grease. He had apparently heard reports of our performance at Thakur Pratap Chand's house. With great difficulty I forced myself to consume three pegs of the evil-smelling liquor, drawing out the interval between successive sips, so that if other considerations failed the mere passage of time should put an end to the drinking session, but it was more than an hour before dinner was served, and my politeness and patience were stretched to their utmost limit. The food was not very palatable, but we were all very hungry and ate large quantities of it. Afterwards we wondered how we could have done it. A radio receiving-set, working off a dry battery, lay on a low table below a garish-looking *thanka*, and seemed to offer good entertainment. We turned it on to hear the nine o'clock news. The announcer told us that the West Indies had beaten us in the third Test Match. The score now stood at 2-1 in their favour.

Altogether it was a disappointing evening.

The next morning we left Gondla at seven o'clock. It was a cloudy day, but the rain kept off, and we were able to maintain a good pace. Once I stopped to take a movie shot of a magnificent glacier across the valley on our right. A mass of shining blue ice lay in a triangular heap on a shelf, high up among dark rugged crags. A hundred feet below it lay another heap of snow and ice in what must be a huge trough, and from here a waterfall started and dropped on to a ledge several hundred feet lower down. The stream ricocheted off the rock and, spreading out into a wide circular fan of spray, disappeared into a slope of rubble. A little lower down it emerged in several branches and went foaming down to the Chandra in a broad and almost vertical delta.

I moved my camera slowly over the scene, and when I had finished, I said to Pal : 'You know, there are glaciers and glaciers. Now if this one had some sense of propriety, it would stage a nice little break of that overhanging cornice, so that I could have a movie shot of the fellow in action.'

I had not finished speaking when I heard a loud roar, resembling the sound of distant thunder. With the quickness of a reflex action I clapped my camera to my eye and set it purring. Through the tiny viewfinder I saw blocks of ice tumbling down from the top shelf, surrounded by a cloud of fine snow. The avalanche ended and the blocks of ice, falling into the trough below, choked the water exit. The flow of the cascade rapidly diminished and in the space of a few seconds ceased altogether. It was a strange spectacle, this sudden death of a huge waterfall, and the incident was accurately recorded by my camera. Every time I project the film I wax eloquent on the topic of well-behaved and considerate glaciers.

We stopped for half an hour at the Sisu rest-house and sat on the lawn sipping a cup of tea. I should have liked to stay longer in the restful and picturesque surroundings of this red-roofed cottage. The river-bed, lying nearly five hundred feet below the lawns of the rest-house, broadens out at this place and offers a vast stretch of fertile land. The fields of barley, rape-seed, *moong* and green peas were in flower and presented a delightful pattern of variegated patchwork as seen from above. Across the river a large waterfall splashed over the rocky face of the mountain in a gigantic triangle of foamy brightness. There was a sense of peace and well-being about everything around us. But we could not tarry and were on our way again at a quarter-past eleven. Another eight miles of dry and undulating route, with half an hour's halt for lunch, in the lee of a tall rock, brought us to Khoksar. We crossed the suspension bridge over the Chandra and climbed up the steep path to the rest-house. Here there was tea and a hot bath, followed by several hours of restful inactivity.

The next day's journey promised to be strenuous ; we had to cover twenty miles and cross the Rohtang pass (13,400 feet), to reach Manali. But it was our last march and our limbs had by now been conditioned to withstand almost any degree of physical exertion. We left our beds at half-past four and were ready at five o'clock. It was decided to postpone breakfast and make a really early start. So after two cups of coffee we set off at a quarter to six. The path went up for two miles at a gentle gradient and then began to climb steeply. I found myself walking with a sprightly step and climbing easily. Pal, who was with me, complained of difficulty in breathing and lagged behind.

I overtook a party of college teachers and fell into conversation with them. They had been as far as Kye-lang and had not ventured to go farther, as they were told that beyond lay only wilderness and the destructive fury of the Himalayas. I spoke to them of the wonder and grandeur of Spiti and made them regret their hasty retreat. This was their first walking tour in the Hima-layas and inexperience was manifest in the way they went up the steep mountain in quick breathless spurts followed by intervals of panting helplessness, when they leaned upon their staves and laughed to see the others doing likewise.

They asked me who I was, and when I told them, they said of course they knew me, they had seen reports of cases heard by me and had read my stories and articles and liked them very much. In return for these compli-ments, I showed them how to waltz up a hill by keeping a slow and even pace, with every third step a little longer than the others : 'Just watch me. *One*, two three, *one* two three, *one*, or if you like, *right* left right, *left* right left, *right* . . .' I explained that this mode of progress was faster and much less tiring because the extra exer-tion of taking a long step was relieved by the two short steps which followed, and since each leg was called upon to do the extra work alternately, there was really an interval of five easy steps after each long one. The accelerated step increased the climber's speed but did not tire him. I also told them about the rhythm of breathing — a quick deep inhalation coinciding with the long step, followed by exhalation during the two short steps. It seemed a little complicated, but the demonstration was easy enough and the theory appealed to their academic minds. They entered into the spirit of the game, and soon I had the whole bunch waltzing merrily up the hill and panting even more vigorously than before.

Shrinagesh rejoined me and we made a dash to the top, arriving there soon after eight o'clock. A cold, piercing wind blew across the pass. There was no shelter in sight and we had to walk the whole length of the pass before we found a rock large enough to protect us from the violence of the gale. Here we sat down near the source of the Beas river and began to eat our breakfast, sausages rolled up in cold *parathas* and a slab of chocolate. Pal was not feeling well and was sick as soon as he had eaten. We put him on a pony and told him to go down as quickly as possible. We knew he would feel better directly he lost a little height.

Shrinagesh and I sat chatting and congratulating ourselves on our achievement — we had done the six miles to the top in just under two hours and a half and were feeling quite fresh. All around us were small and large slabs of stone propped up to stand erect. There were hundreds of them, giving the scene an appearance

of a dead forest with small withered stumps growing
out of the barren land. The stones had been erected by
visitors to the Beas *kund*[1] as an offering to the gods on
behalf of themselves and their families — one for each
person. It was a cloudy day with low mist enveloping
the pass and we were denied a view of the snow-range
and the double Gyephang peak in Lahoul which present
a magnificent sight on a clear day. I asked Shrinagesh
if it was worth while climbing up another thousand
feet to see Akbar's lake.

'Akbar's lake?' he asked.

I expressed surprise. 'Don't you know? Well,
then, lean back and listen. The great Mogul Emperor,
Akbar, had once a daughter born to him. She was a
most lovely creature, but one of her legs was deformed
and shrunken like a withered, sapless branch. Physicians
and sages were summoned from all corners of his
kingdom, but no one could effect a cure or discover the
cause of the piteous affliction. The Emperor and his
consort were at their wits' end. Day by day the young
princess grew fairer, but her beauty was marred by the
hideous limb adhering lifeless to her shapely body. The
queen wept and sorrowed, the Emperor went about the
business of his empire with a solemn face. Astrologers
and soothsayers, divines and pandits were called, but
their science and their holiness were of no avail. Finally
from the far Himalayas came an old yogi carrying a
staff in one hand and a vessel or gourd in the other. His
body was smeared with ashes, and silver locks hung
below his neck. He heard the whole story from the

[1] *Kund* — pool or spring. The Beas *kund* marks the source of the
Beas river.

lips of Akbar himself and, after meditating for a while,
spoke in these terms :

'"Far, far from here, but within the boundary of
your kingdom, and in the very heart of the Himalayas
is a lake at a great height. It is full of limpid blue water
and is continuously gazed upon by the snowy peaks of
the Himalayas ; but seldom is it visited by man or beast.
Shortly before your daughter was born a mare lost her
way and strayed to the bank of the lake. In trying to
quench her thirst she slipped and fell into the water and
was quickly drowned. But in falling, she turned over
on her back and one of her legs sticks out of the lake
like a horrible protuberance. The gods have taken
offence, and this ugly sight is a constant reminder of your
negligence and the forgetfulness of your officers. Unless
the mare's body is completely sunk and the leg dis-
appears from view, the princess will not recover."

'The yogi went back to his solitary abode in the
mountains, and the Emperor sent a thousand men to
scour the Himalayas from one end to the other, and
discover the lake with the mare's leg. They had instructions to push the dead limb into the water till it was
completely submerged, and make sure that no untoward
accident would expose it to view again. They were
to note the day and the hour when the task was
completed.

'Days of anxious waiting passed, but the Emperor
could see no change in the condition of the young
princess. The hopes awakened by the ancient yogi
were once again buried in gloom. Then quite suddenly,
one day, as Akbar was sitting by his daughter's cradle
and watching her baby movements with a heavy heart,

he noticed something different. The withered leg seemed to have altered in appearance and he thought he saw a slight tremor passing over it. With a cry of joy he lifted his baby daughter in his arms and ran his hand over the leg. It felt warm and alive to his touch. From that moment it recovered rapidly, and in the space of a few days the dead limb was completely restored to a normal state.

'A month later a party of the Emperor's men returned and informed him that on a certain day, at a certain hour they had chanced upon a lake situated above a pass at a great height and surrounded by snowy peaks. In the lake they had seen an evil-looking object which, on approaching, they recognized to be the leg of a horse. They pushed the leg in and weighted it down with rocks. They noted the day and the hour accurately and returned to the capital post-haste. The time mentioned by them corresponded exactly with the moment at which the Emperor had observed the beginning of life in his daughter's withered leg.

'That lake,' I concluded, 'lies just above the hill in front of us, and I have heard it said that the body of the dead mare still lies below the surface of the water.'

Shrinagesh stood up and brushed the seat of his trousers with his hands. 'That is a charming story,' he said, 'but I think on a day like this Akbar's lake may be taken as read. We have done quite enough climbing for the day.'

The down-hill path was a paved stoneway descending in an interminable series of steps. It was a crazy pavement gone completely crazy. We careered down, turning and twisting round the sharp bends with the speed of horses returning home after a long outing. Our knees trembled and the tendons round the knee-cap sprang into consciousness at every step but we kept on moving. We were near the end. At Manali there would be time enough to relax and rest our limbs. Major Banon had asked us to dine with him. He would serve Scotch whisky and a meal that would put the Lord Mayor's banquet to shame.

We crossed little streams and snow-bridges. The Raini *nala* [1] was buried in a vast stretch of snow over which the P.W.D. labour had carved out a picturesque, though wholly unnecessary, path. We walked along it, but the mules could not be persuaded to deviate from the route their leader had picked out, straight across the snow slope. It was perfectly safe because the surface of the snow was coarse and the gradient not too steep to permit an easy passage almost anywhere.

The bottom of the valley came into view. Even from a distance it seemed filled with peace and beauty. The slopes leading down to the banks of the Beas were covered with a carpet of deep-green turf across which ran an elusive pattern of buttercups and daisies, anemones and dandelions, with here and there a splash of wild primulas. The river lay like a bright ribbon of silk, playing hide-and-seek among cedars and Himalayan oaks. On the left a waterfall dropped four hundred feet from the top of a cliff and disappeared from view. To our right stood a massive wall of smooth bare rock, its face streaked by the flow of a strange inky liquid which in ages past must have been squeezed out by the force which pulled the mountain out of the bowels of

---

[1] *Nala* — stream.

the earth. At the foot of this wall a stream rushed out of a narrow crevice and plunged into the Beas. The valley was deep and narrow, but it was sheltered from the gale which blows perpetually a thousand feet above, and it was green — a beautiful comforting green that spoke to you softly of the fecundity of Mother Earth, and welcomed you to her warm caressing embrace.

We left the path to pay a visit to the Rahalla Falls, cunningly concealed behind a small spur. Three large streams of water came rushing over their rocky beds from three different directions, converging towards the same point, and just before the moment of meeting flew over the rocks in three gigantic cascades, roaring and splashing into a cauldron fifty feet below. The monsoon had been active for nearly a month, and the volume of water had swollen to its maximum limit. It was a magnificent sight and we were reluctant to leave it.

We climbed down the rocks and walked for some time over the soft springy turf along the bank of the River Beas. Rahalla is a charming little spot nestling in a corner of the greenest of green valleys, protected by mountains which rise sheer to a height of several thousand feet on either side. But there is no feeling of being enclosed or imprisoned ; instead there is a sense of calm tranquillity, comfort and well-being. In the old days there used to be a P.W.D. hut near a rustic bridge, where travellers could stay, but some years ago the hut was accidentally burnt down, and war-time stringency prevented its being rebuilt. An enterprising man from the neighbouring village of Kothi has put up a mean-looking shelter by erecting a wall of stones against a natural cave in the hillside. A wooden board nailed to a tall post proclaims that this is THE NATIONAL HOTEL RAHALLA. Here tea and biscuits are served. Tea is two annas a cup, and biscuits are an anna apiece. The place was black with smoke and reeked of the urine of mules. We decided to go on to Kothi, only two miles farther, before seeking refreshment, and, crossing the bridge over the Beas, continued on our way through the beautiful green valley.

Half a mile beyond, a mule-man pointed to a small hole in the hillside and asked me to place my hand over it. A cold draught blowing out of the rat-hole chilled my hand. There were several such holes in the space of a hundred yards along the path, some exhaling faintly and some giving forth a lusty blast to the accompaniment of a low moan. The mule-men told us that on a stormy day these little mouths blew so hard and whistled so loudly that they struck terror into the stoutest heart. There is no mention of these air-holes in the local gazetteer nor in any of the monographs dealing with the geological peculiarities of the region, and I would have liked to discover the end of the subterranean pipe of which the existence was disclosed by each little opening. The other ends must be a long way up, on the very top of the mountain where they catch the cold air of the higher strata and, sucking it in, blow it out of the lower end in a strange and eerie manner. But the top of the mountain was not within our climbing capacity, nor could we spare the time for this adventure. So gathering the wonder of one more thing that happens in the Himalayas, we proceeded on our way to Kothi.

At the rest-house we rested and lunched. The *chow-kidar* gave us cow's milk, fresh, wholesome and palatable. We drank copiously of it, and sat in the verandah looking at the high mountain across the valley. It rose sharply to an altitude of twelve or thirteen thousand feet, and, perched precariously on its steep, grassy face, goats and sheep were grazing. At nightfall they would be rounded up, collected in a heap and stored in a small grassy pocket in front of an overhanging rock which provides the night's shelter for the *gaddis*. To our right the valley disappeared round a bend and wound its way to the Rohtang pass.

Two Englishwomen, very plain and very talkative, came and began to speak about themselves. They were sisters — in India for a short holiday — one was a barrister, a very promising one, only she had ruined all her prospects by contracting matrimony. She referred to this unfortunate event as if it were some fell disease which had attacked her in an unguarded moment and destroyed her faculties and all her strength and vitality ; the other was an artist — an amateur painter — only 'unfortunately all my paintings have been bought', so like the poor dog she had none. They gushed and fussed even more when we answered their queries about our identity and the sphere of our activities. They were very surprised — yes, amazed, said the other one — that we had not, after independence, changed the names of streets and houses and discarded the foreign element — Kingsway, Queensway, Connaught Circus. Delhi was simply littered with the remnants of British Raj, and wouldn't it be more patriotic to have native names — or shouldn't she use that word, it *has* a horrid sound

— why, even the houses in Simla where they had gone for a brief week-end — too brief unfortunately — for Simla was *so* beautiful — *all* the houses had English names — well, yes, it might be a little confusing at first, but she had *no* doubt at *all* that it would be more consistent with the dignity of a *free* country — of course it was very flattering to them — very gratifying indeed to . . .

'Well, good-bye, *good*-bye, it has been terribly thrilling meeting somebody civilized like this, don't you think, good-*bye* and *good* luck, good *luck*.'

We went down to the wooden bridge over the Beas and gazed into the dark and narrow chasm more than a hundred feet deep, at the bottom of which the river twists and groans like an angry serpent, foaming at the mouth, hurtling over the steep bed of stones, rocking from side to side, bouncing off the scarped rocks as if it were writhing in pain on being confined within such narrow limits. The gorge, at this place, is only a few feet wide, and at some points you could stretch out your arms and touch both sides with your finger-tips. We were told that an American tourist once did this. He had himself lowered in a large basket hung over the railing of the bridge, and went all the way to the bottom touching the walls of the gorge with his outstretched hands.

We passed through Kothi village, a picturesque collection of flat-roofed houses lying along the fold of the mountain and overlooking the gradually widening valley of the Beas. Once again we crossed the river and passed through a succession of pine groves. At one place a massive wall of rock several hundred feet high

stood a little away from the path to our left. Its face
was black with the ravages of wind and rain, but near the
top was a large clean patch, thirty feet long and ten feet
wide, as if a large tablet of stone had been neatly sliced
off with a monstrous carving-knife. One of the
muleteers saw me looking up at it and told me the
following story :

Many many years ago, the young daughter of the
ruler of Kothi was married to the Prince of Nagar and
had to go a long way from home. She languished for
a sight of her native soil and sent a message to her
father asking him for a clod of earth or a piece of rock,
something she could touch and put her foot on, something that would charm away the ache of nostalgia in
her heart. The Raja received the message and was
filled with concern for his daughter's happiness. Some
bees who were collecting honey from the flowers in
the palace also heard the message. They went back to
the hive and told everyone that their princess was unhappy. On this the entire swarm left the business of
honey-making and began to work on the face of the
huge rock. Soon they had cut away a slab several
yards in length, and this they carried and laid before
their princess in Nagar. The hum of the bees was
heard throughout the land, and the people were glad,
for now the young princess would be able to sit, or
walk, or lie down on a piece of her own native soil
whenever the desire arose within her.

The muleteer added that the slab of rock lies near
the Nagar village to the present day, and he assured me
that many curious visitors had measured the dimensions
of the slab and compared its size and shape with those

of the white patch on the parent rock. They had gone away amazed but completely convinced.

Two miles from home Bachittar Singh's jeep was waiting for us, like a surprise packet. He had ridden home and sent it to meet us. We drove along the narrow uneven path, bumping over stones and boulders, and came to the cantilever bridge whose quaint construction had delighted me on the first day of our trek. The planks groaned and creaked, but took the weight of the loaded jeep. Then up the steep slope with the engine grinding stridently, while the driver murmured 'four by four', and played with the gear knobs, coaxing them back and forth. Past the post-office, the sweet-shop and the *dak* bungalow. Along the stone-paved lane and through the avenue of lime trees leading to the Forest rest-house.

The jeep ceased its groaning ; we jumped out and saw a tree laden with purple plums bursting with luscious ripeness. Near it was a tree with large pale-green pears hanging from every branch, and another with red and green apples. We were back in the Kulu valley — the valley of peace and plenty.

After tea I sat on the verandah writing the last chapter of my diary. There was a strange and compelling quiet in the stillness of the landscape beyond the rest-house. The mules had, long since, unloaded their burden and carried away the tinkling of their bells. A small water-course murmured its way to the experimental nursery of the Forest Department. At the end of the lawn a row of hollyhocks stood still and erect. Beyond, in the far distance, dark, pine-covered hills sloped down from opposite sides to form a large V against a cloudy sky.

In the distance also flowed the Beas, and if I listened intently I could hear the dull and muffled groan of its torrent. In the sad, restful ease and hush of the July afternoon a sense of soft drowsiness enveloped me ; but from time to time, as scenes of sun and snow and purple crags flashed across my half-closed eyes, my mind was teased by a desire to go back to the real mountains far away, beyond the high passes and among the cruel rugged rocks where the barren splendour of the Himalayas reigns supreme.

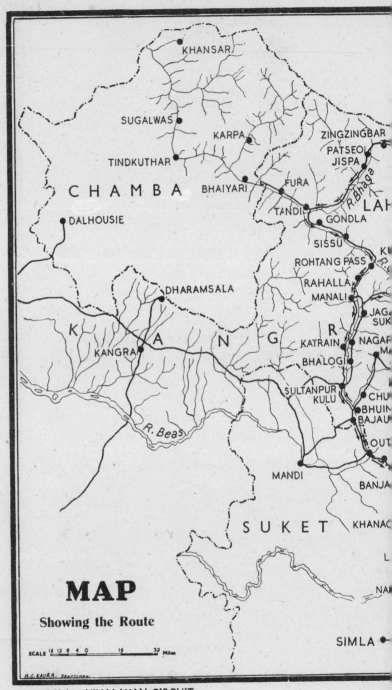

MAP

Showing the Route

SCALE 16 12 8 4 0   16   32 Miles

H.C.KAURA, DRAFTSMAN.

KHOSLA : HIMALAYAN CIRCUIT